NATIVE
New Zealand
FLOWERING
PLANTS

NATIVE

New Zealand
FLOWERING
PLANTS

J. T. Salmon

REED

Dedicated to all those fine people who work
daily to preserve our environment and who
care deeply for the earth and all its living
things.

Cover photographs (left to right): kowhai flowers (fig. 593): flowers of
rata vine (fig. 572); drupes of the karaka tree (fig. 50).

The verso of the half title page shows heketara flowering in scrub
(fig. 231); the title page shows flowers of akepiro (fig. 226). Page
vii shows a southern rata in flower, Haast River, January.

Published by Reed Books, a division of Reed Publishing (NZ)
Ltd, 39 Rawene Road, Birkenhead, Auckland. Associated companies,
branches and representatives throughout the world.

ISBN 0 7900 0223 X

© 1991 John T. Salmon

The author asserts his moral rights in the work.

First published 1991
This edition 1999

Printed in Hong Kong

CONTENTS

PREFACE

This book started out as a revision of my previous book *New Zealand Flowers and Plants in Colour*, published in 1963 and since reprinted eight times. First, the idea was to make it a 'field guide' but it soon became clear that a more comprehensive production was desirable and, indeed, possible. This new volume represents about forty years of exploring the New Zealand wilderness, whenever opportunity presented itself, photographing our native plants — at first by myself, but later with my wife and sons. The wild places and their plants became the centres of family holidays and the photographs gathered the backbone for future writings. In all this I have received help from many people who have been acknowledged in my earlier books, but for this one I have been wonderfully supported by my wife, Pam, not only when we went into the wilderness areas but most especially during the writing and preparation of the book. She has toiled unsparingly as a proof-reader and as a critic, offering many useful suggestions as we went along, and for all of this I am really grateful.

J. T. Salmon
Taupo
December 1990

INTRODUCTION

Within the small zone encompassing the islands of New Zealand, early European explorers found a flora that proved to be one of the most remarkable on earth. In the lowlands or in the mountains, by coast, stream, river or lakeside, plants of all kinds grew in unlimited profusion. Though the number of plant species found in our islands proved ultimately to be less than was at first predicted, few other regions have possessed such a high proportion of plant species not found anywhere else. Among flowering plants alone, 75 percent of the species proved peculiar to New Zealand, and few other regions could show within so small an area such an infinite variety of habitat, from seaside to alpine situations.

Into this near pristine environment with its lovely world of plants, about 150 years ago, came Europeans with their axes and saws, animals and fire. The early settlers saw the forest as a hindrance to farming, grass, crops and wealth. So the frontiers of the 'bush', as the forests were called, were pushed back towards and into the mountains, while introduced game animals increased the havoc being wrought upon the land.

From the earliest European pioneering days New Zealand society developed a spirit of indifference towards our native plants, which has changed for the better only within the last thirty years. Meanwhile many New Zealand native plants have been cultivated and appreciated in such other countries of the world as Australia, California, Britain and Spain.

I hope that this book will help to enhance the public appreciation of our natural heritage of native plants, many of which are unique, too many of which are now threatened with extinction, and almost all of which are beautiful to behold.

This book of pictures is designed to show the flowers and fruits of the native plants of New Zealand most likely to be met with when exploring the countryside, and to help interested people to identify our native plants. It moves from seaside plants through scrub and forests to the open spaces of the high mountains, grouping the photos according to the habitats in which the plants grow. However, though many plants can grow in more than one type of habitat, each is shown only once, usually in the habitat in which it is commonly found. The short text accompanying each picture should, with the picture, provide sufficient detail to accurately identify each plant.

The flowering times given in the legends and the months in which each photograph was taken (as recorded in the captions) are a general guide only, as flowering times can vary by as much as three to four weeks from place to place, depending on latitude, elevation, situation and seasonal climate.

To identify a particular plant, look first in that section of the book that corresponds to the natural habitat in which you were when you found it. A comprehensive index at the end of the book will assist in tracking down plant names. A list of reference works is appended for those readers who wish to learn more about our native flora.

NORTH ISLAND

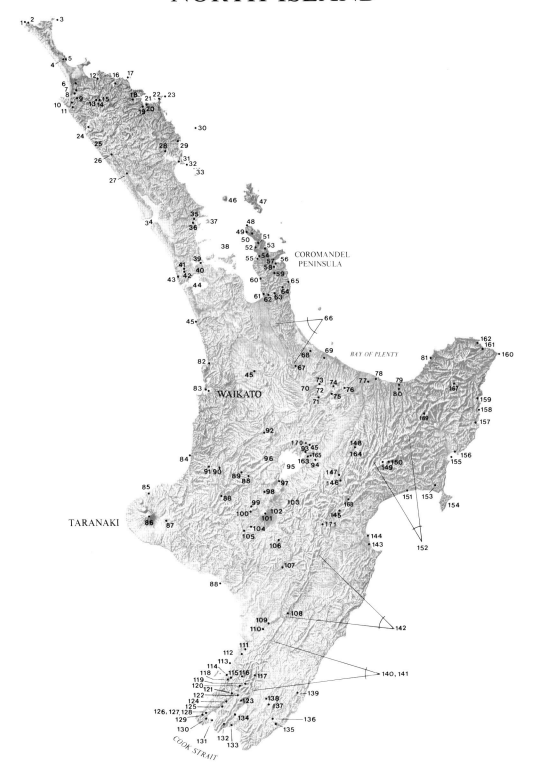

•O

1 •2   •3

•4 •5

12   16   17

6   18   22 23
7 8   21
10 13 14 15   19 20
11

•30

24
28   29
25   31
26   •32
27   33

•46   •47

35   37   48
34   36   49   51
50   53
38   52
55   54 56   COROMANDEL
57   PENINSULA
58
39   59
41   40
42   65
43   44   60   64
61   62 63

•45   66   69

82   68   BAY OF PLENTY   81
67   162
45   161
83   WAIKATO   73 74   78   160
70   72   76 77   79
71   75   80   167

92   159
158
157
170
93 45   148
163 94 165   164   169
84   95   156
91 90   89   147   150   155
88   146   149
97
98   151   153
88   99   103   168   154
100   102   145
85   101   171
104   144
86 87   105   143
106   152

107

88•

108   142

109
110

111

112
113
114   115 116
118   117
119   140, 141
120 121
122   139
124   123   138
125   137
126, 127, 128
129   134
130   136
131   132   135
133

COOK STRAIT

SEASIDE PLANTS

Seaside plants are those found growing on the rocks, cliffs, sand-hills and sandy beaches, marshes and swamps within close proximity to the sea. They are all hardy plants adapted to withstand strong winds, shifting sands and frequent drenching with salty spray. Let's look first at those among the rocks and cliffs.

1 Pohutukawa flowers, Coromandel Peninsula (January)

2 The hairy seeds of pohutukawa, Karaka Bay, Wellington (July)

1–2 Pohutukawa/New Zealand Christmas tree, *Metrosideros excelsa*, grows naturally around Auckland, the coasts of Northland, Coromandel Peninsula, Bay of Plenty and East Cape as a medium-sized, rounded tree or a massive, often twisted and gnarled, large, spreading tree. During December and January it covers itself with masses of deep crimson to blood-red flowers (fig. 1). The conspicuous part of the flower is the mass of red stamens that surrounds a calyx funnel filled with nectar, which is much sought after by birds. The conspicuous hairy seed capsules (fig. 2) split open during June–July to release many seeds but seedlings are seldom found abundantly.   MYRTACEAE

3 Kermadec pohutukawa, *Metrosideros kermadecensis*, is similar to the familiar pohutukawa, *M. excelsa*, but is a smaller tree with shorter, oval-shaped leaves, 2.5 cm long by 10–20 mm wide. Flowers occur throughout most of the year and, though native to the Kermadec Islands, the tree is grown extensively in parks and gardens in New Zealand.   MYRTACEAE

3 Kermadec Island pohutukawa flowers, showing also the typical rounded leaves of this species (January)

4 ***Pyrrosia serpens*** is a peculiar spore-bearing plant found commonly in coastal rocky places but it also grows as an epiphyte on trees in lowland and montane forests throughout New Zealand, the Kermadec and Chatham Islands.

POLYPODIACEAE

4 Plant of *Pyrrosia serpens,* showing spores, Wellington Harbour (January)

5–7 Taupata/angiangi/naupata, *Coprosma repens*, grows along our coasts, south from North Cape to Marlborough, in rocky or sandy situations and, with its shining, bright green leaves, is one of the best known and most beautiful of our coastal shrubs. Flowers occur during October and November (fig. 5, female; fig.6, male), with the 6 mm long drupes (fig. 7) colouring in March and April.

RUBIACEAE

7   Taupata berries, Wellington (January)

8 Woollyhead, *Craspedia uniflora* var. *maritima*, grows in coastal rock crevices and grassy places around Cook Strait and near Ocean Beach at Oamaru.   ASTERACEAE

5   Taupata, female flowers, Wellington (October)

6   Taupata, male flowers, Wellington (October)

8   Woollyhead in flower, Cook Strait (November)

9   Horokaka in flower, Cook Strait (December)

10   Thick-leaved porcupine plant showing berries, leaves and stems, Wellington hills (February)

9   Horokaka/Maori ice plant, *Disphyma australe,* is found along the coasts of New Zealand as well as the Kermadec and Chatham Islands, trailing on rocky cliffs, coastal banks and gravels. The flowers, 2.5 cm across, occur from October to March.

                                     AIZOACEAE

10   Thick-leaved porcupine plant, *Melicytus crassifolius,* occurs along our coasts as large, flat, springy cushions, up to 20 cm thick, closely adpressed to rocky surfaces. The flowers are only 3 mm across and appear from September to January, and the berries, 6 mm across, are ripe from October to March.   VIOLACEAE

11   Wild celery, *Apium australe,* occurs all along New Zealand coasts in rocky places and flowers from December to January. It is also found on the Kermadec, Three Kings and Chatham Islands.

                                     APIACEAE

11   Wild celery in flower, East Wairarapa coast (December)

12 Wharanui showing leaves and
a flower-spike, southern
Wairarapa coast (August)

14–15 Large-leaved porcupine plant, *Melicytus
obovatus*, is an erect, spreading shrub up to 3 m
high, with hairy young branchlets. It grows in rocky
coastal places throughout both the North and South
Islands. The thick adult leaves are simple, 4 cm long
by 15 mm wide (fig. 14), but juvenile leaves are
toothed and lobed. Small solitary flowers occur
during November and December; the berries (fig.
15), 3–5 mm across, ripen from March to December.
VIOLACEAE

12 **Wharanui,** *Peperomia urvilleana*, is a succulent,
prostrate and branching herb up to 30 cm high,
found on rocky coasts near the sea throughout the
North Island and the northern tip of the South
Island. The leaves are about 4 cm long by 2 cm wide,
with petioles up to 5 mm long; the flower-spike is
up to 5 cm long. The plant flowers and fruits
throughout the year. PIPERACEAE

13 **Pigweed,** *Rhagodia triandra*, grows as large
mats often hanging over rocky faces beside the sea,
from North Cape south to about Timaru in the east
and Jackson Bay in the west, and also on the Kermadec Islands. Minute flowers occur from September to November, and the berries, 5 mm across,
ripen from December to March.

CHENOPODIACEAE

13 Pigweed with berries, Wellington coast
(December)

14 Large-leaved porcupine plant,
showing leaves and berries,
Cook Strait (December)

15 Large-leaved porcupine plant
heavy with fruit, Cook Strait
(December)

17   Creeping fuchsia with berries, Wellington (March)

16   Creeping fuchsia with flowers, Wellington (November)

16–17 Creeping fuchsia, *Fuchsia procumbens*, forms a sprawling or trailing shrub found along North Island coasts from North Cape to about Thames in stony, rocky, sandy and gravelly places above normal high-tide mark. Flowers, about 18 mm long, with blue pollen (fig. 16) occur from October to January, and the berries (fig. 17) ripen from January through March.   ONAGRACEAE

18   Cook Strait Islands senecio in flower, Wellington Harbour (January)

18   Cook Strait Islands senecio, *Senecio sterquilinus*, is a plant up to about 1 m high, rather similar to *S. lautus*. It is found growing in rocky places and sea-bird nesting areas on the islands of Cook Strait only.   ASTERACEAE

19   Shore groundsel, *Senecio lautus*, is a much-branched, herbaceous plant that grows commonly in rocky places, sometimes on sand dunes, near the sea but can be found also in inland rocky places up to 1,500 m altitude. Flowers 10–25 mm across occur during November and December.   ASTERACEAE

19   The shore groundsel in flower, Karaka Bay, Wellington (November)

20-21 *Colobanthus muelleri* is one of 15 species of *Colobanthus* found in New Zealand. Others occur in South America, Australia, Tasmania and some subantarctic islands. *C. muelleri* is found along the coasts of the North Island and the Chatham Islands in gravelly places, forming flat rosettes 10–20 mm across with leaves 10-15 mm long (fig. 20). Flowers and fruit capsules each 4–6 mm long (fig. 21) appear from September till April.

CARYOPHYLLACEAE

22 Pohuehue with flowers and seeds, Makara coast (April)

22 Pohuehue, *Muehlenbeckia complexa*, grows as a dense, tangled mass, several metres across and up to 60 cm high, all along our rocky coasts as well as inland in coastal and montane forests, where it can cover low shrubs with impunity. It flowers profusely from October to June, producing black seeds sitting in the white, cup-like tepals. POLYGONACEAE

20 *Colobanthus muelleri* plant with fruiting capsules, Makara coast (November)

21 Fruiting and flowering capsules of *C. muelleri*, Makara coast (November)

23 Sprawling pohuehue, *Muehlenbeckia ephedrioides*, forms a much-branching, sprawling shrub, which bears small flowers from November to June. It is found all over the North and South Islands in coastal scrub and in rocky, sandy and gravelly places from sea-level to 1,000 m. It is similar to *M. complexa* except that the rigid, wiry stems have only a few narrow leaves 5–25 mm long.

POLYGONACEAE

23 Stem of sprawling pohuehue showing seeds, Wairarapa coast (February)

Turning to seaside plants that grow in sandy and gravelly places, we find:

24 Flowers of mutton bird sedge, male above, female below, Stewart Island (October)

24 Mutton bird sedge, *Carex trifida*, has a characteristic spike with the long, dark brown male flower above the shorter, lighter-coloured female flower. This sedge is found on coastal cliffs and rocky places as well as in inland swamps south of Timaru, on the subantarctic islands, the Chatham Islands and in Chile. It forms dense clumps about 60 cm across and has keeled leaves, 10 mm wide and 1–2 m long, with rough or scabrous margins. CYPERACEAE

25 Pingao in flower, Tautuku Beach (December)

25 Pingao, *Desmoschoenus spiralis*, is a sedge found covering large areas of sandhills along the New Zealand coasts. Flowers and seed-heads are produced during December and January.
CYPERACEAE

26 Sand-dune coprosma, *Coprosma acerosa*, forms a low-growing, cushiony mass of interlacing branches with narrow leaves, 1–5 mm wide and 12 mm long. Tiny axillary flowers are followed by drupes, 7 mm long, which turn pale blue during March and April.
RUBIACEAE

26 Sand-dune coprosma plant with drupes, Tautuku Beach (January)

27 Sand-dune pin cushion, *Cotula trailii*, with its conspicuous dark, hairy flower-stems, is found sparsely among sand dunes, on sandy beach terraces and on salt marshes from Cape Foulwind south to Stewart Island. Flowers occur from December till February. ASTERACEAE

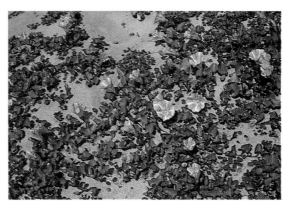

29 Nihinihi plant flowering and spreading over sand at Tautuku Beach (January)

27 Sand-dune pin cushion showing flowers and leaves, Stewart Island (December)

28 Spiny rolling grass, *Spinifex hirsutus*, occurs as a sand-binding plant that stabilises sand dunes along the coasts of the North Island and the northern parts of the South Island. Not indigenous, it is also found in Australia, Tasmania, some Pacific islands and India. GRAMINEAE

28 Spiny rolling grass with flowers and seed-heads, Warehou Bay (December)

30 Nihinihi flowers, Tautuku Beach (January)

29–30 Nihinihi/sand convolvulus, *Calystegia soldanella*, is found throughout New Zealand, growing on sand dunes, sandy beaches (fig. 29) by sea or lake shore, gravelly riverbanks and occasionally on rocky places. Flowers (fig. 30), 2.5–7.5 cm across, occur in profusion from December to March.

CONVOLVULACEAE

Seaside plants that grow in sandy and swampy places or salty marshes include:

31 These plants of the **seaside daisy,** *Celmisia major* var. *brevis*, shown here in flower, were found at Wharariki Beach, near Farewell Spit, growing in the sand and on rocky ledges close to high-water mark, where they are frequently drenched by salt spray. They are similar to *C. major* var. *brevis* from Mt Taranaki but are much smaller.

ASTERACEAE

32 Triangular-stemmed sedge in flower, Farewell Spit (January)

31 Seaside daisy plants in flower on Wharariki Beach, near Farewell Spit (January)

33 Triangular-stemmed sedge growing near Farewell Spit (January)

32–33 Triangular-stemmed sedge, *Scirpus americanus*, is commonly found in damp, sandy places, sometimes covering extensive areas, along the New Zealand coasts from the Coromandel Peninsula southwards but is absent from Westland and Fiordland. It flowers (fig. 32) during November–February.

CYPERACEAE

34 Sand sedge flower-head, South Wairarapa coast (November)

34 Sand sedge, *Carex pumila*, is a coarse, tufted sedge that grows along our coasts on the seaward slopes of sandy and gravelly beaches. It is not indigenous and is found in many places round the Pacific Basin.

CYPERACEAE

35 Maori musk, *Mimulus repens*, is found throughout New Zealand, growing in coastal salt marshes or shallow swamps near the coast. The musk-scented flowers, which occur from October till January, when first open are a deep mauve colour as shown but, after a few days, they fade to white.

SCROPHULARIACEAE

35 Maori musk in flower, Lake Wairarapa outlet (December)

36 Remuremu, *Selliera radicans*, is a perennial that grows on mud-flats, coastal, damp, sandy and rocky places or inland along stream, pond or lake margins. Flowers, usually white but sometimes pale blue, occur freely from November to April.

GOODENIACEAE

36 Remuremu plant in flower, Karaka Bay, Wellington (February)

37 Flowers of maakoako, Wellington Harbour (April)

37 Maakoako, *Samolus repens*, is a creeping perennial that grows in salty marshes and rocky places beside the sea throughout New Zealand. Flowers arise at the branch tips as well as from the axils of the thick leaves and occur from December till April.

PRIMULACEAE

38 Yellow buttons, *Cotula coronopifolia*, is a creeping plant found along the edges of coastal muddy swamps, in damp sand-dune hollows and along damp lowland streamsides. The flowers, about 8 mm across, arise from October to January, but the seeds do not ripen until the following November.

ASTERACEAE

38 Yellow buttons plant in flower, Wellington (October)

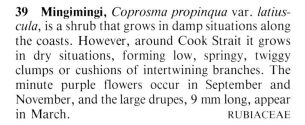

39   Mingimingi plant showing leaves and drupes, Cook Strait (March)

40   Southern salt horn plant growing on rock shelf at Karaka Bay, Wellington (February)

39 Mingimingi, *Coprosma propinqua* var. *latius-cula*, is a shrub that grows in damp situations along the coasts. However, around Cook Strait it grows in dry situations, forming low, springy, twiggy clumps or cushions of intertwining branches. The minute purple flowers occur in September and November, and the large drupes, 9 mm long, appear in March.   RUBIACEAE

40–41 Southern salt horn, *Salicornia australis*, is a prostrate, herbaceous plant that grows in salty marshes, and rocky crevices or on stony beaches near high-tide mark, throughout New Zealand (fig. 40). Fig. 41 shows the minute flowers that arise on the upper parts of the stems from November to March. Fruits, 2 mm across, occur from December to April.   CHENOPODIACEAE

41   Close view of southern salt horn flowers, Karaka Bay, Wellington (January)

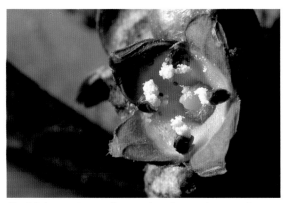

42   Mangrove flower and flower-bud showing pollen, Whangaparaoa (April)

43   Male mangrove flower showing anthers ready to discharge pollen, Whangaparaoa (April)

44   Mangrove flowers showing (left) perfect anthers
     and (right) anthers after pollen has been
     discharged, Whangaparaoa (April)

45   A perfect mangrove flower showing female
     stigma, Whangaparaoa (April)

46   A mature mangrove tree with
     seeds near Paihia (February)

42–48   Manawa/New Zealand mangrove, *Avicennia
marina* var. *resenifera*, is a medium-size tree or shrub
that grows, except at low tide, with its roots covered
by salt water (fig. 46). Because of this, structures
called aerial roots grow upwards (fig. 48) to allow
the main roots to breathe air when uncovered by the
tides. Flowers are 6–7 mm across; the female flower
(fig. 45), with its broad pistil, arises as 4–8-flowered
clusters from February to April, but the fruit (fig.
47) takes until the following January to ripen.

AVICENNIACEAE

47   Mangrove seeds, Russell (December)

48   Aerial roots exposed at low tide. These allow the
     main root system of the mangrove to breathe,
     Tapatupoto Estuary, Northland (January)

COASTAL PLANTS

Trees, shrubs and herbaceous plants found growing along the coastal strips and hills, around river mouths and estuaries, on cliffs and rocky places or in bogs and swamps, near the coast but not beside the sea, are our coastal plants. Some unusual forms of these can range far inland as well and high into the mountains.

49 Flowers of the karaka tree, Karaka Bay, Wellington (October)

50 Drupes of the karaka tree, Karaka Bay, Wellington (March)

51 Close-up of karaka flowers, Karaka Bay, Wellington (October)

49–51 Karaka, *Corynocarpus laevigatus*, is a large tree, up to 16 m high, with large, dark green, very glossy leaves. It grows as isolated specimens or in small groves in coastal districts of the North Island, and in the South Island as far south as Banks Peninsula in the east and Jackson Bay in the west. The flowers are borne on stiff panicles (fig. 49), up to 22 cm long, from August to November, with the individual flowers (fig. 51) each about 5 mm across. The drupes (fig. 50) ripen from January through April and are 2.5–4.3 cm long.

CORYNOCARPACEAE

52 Rauhuia, *Linum monogynum*, is a bushy herb found throughout New Zealand. The flowers are up to' 3.2 cm across and appear from October to December, though in warmer districts they may continue to April.   LINACEAE

52 Rauhuia flowers, Karaka Bay, Wellington (March)

54–55 Napuka, *Hebe speciosa*, is a lovely evergreen shrub, having either purple (fig. 55) or magenta (fig. 54) flowers. Both colour forms are found growing naturally on western coastal cliffs of the North Island from Hokianga to Tongaporutu and Cook Strait, and in Pelorus Sound; it is now often grown in private gardens. The flower-spikes, up to 5 cm long, arise from January through winter till October.

SCROPHULARIACEAE

53   Tauhinu spray with flowers, Makara coast (April)

53 Tauhinu/cottonwood, *Cassinia leptophylla*, grows plentifully in coastal areas from East Cape to Nelson, bearing clusters of flower-heads, up to 2 cm across, from November to January and again during March and April.   ASTERACEAE

54   Napuka with magenta flowers, Cook Strait (October)

55   Napuka with purple flowers, Cook Strait (February)

56 Northern koromiko, *Hebe obtusata*, with its pale lilac-coloured flower-spikes, up to 6 cm long, is found along the coast from Manukau Heads to Muriwai Beach. It flowers from January till June.

SCROPHULARIACEAE

56   Flowers of northern koromiko, Manukau Heads (April)

57　Spray of flowers of coastal tree daisy, Karaka Bay, Wellington (April)

57–58　Coastal tree daisy, *Olearia solandri*, grows on coastal hills and rocky cliffs near the sea, up to 500 m altitude, throughout the North Island and in coastal areas of Marlborough and Nelson in the South Island. The flowers (fig. 57), up to 10 mm across, appear from February through to April, and the fluffy seeds (fig. 58) remain until October.

ASTERACEAE

58　Seeds of coastal tree daisy, Karaka Bay, Wellington (September)

59　Rengarenga in flower, Waikanae (December)

59　Rengarenga, *Arthropodium cirratum*, is a handsome lily found growing in dry, rocky, coastal regions of the North Island, Nelson and Marlborough. The plant flowers abundantly during November and December.　ASPHODELACEAE

60　Green clematis, *Clematis hookeriana*, grows in dry places along both the north and south shores of Cook Strait. Sprawling over rocks and shrubs, it produces its green or pale yellow, sweet-scented flowers, about 15 mm across, from November till January.　RANUNCULACEAE

60　Flower spray of green clematis, Cook Strait (November)

61 Flowers and leaves of Castlepoint groundsel, Castlepoint (December)

62 Cook Strait groundsel in flower, Cook Strait (December)

61 Castlepoint groundsel, *Brachyglottis compacta,* grows naturally only on limestone cliffs near Castlepoint and flowers profusely from December to February. ASTERACEAE

62 Cook Strait groundsel, *Brachyglottis greyi,* is a handsome, grey-leaved shrub restricted in the wild state to rocky places on the north shores of Cook Strait and northwards to the Pahaoa River. The bright yellow corymbs of flowers, each flower about 2.5–3 cm across, occur from December till March. ASTERACEAE

63 Wharariki flowers, Karaka Bay, Wellington (November)

64 Vittadinia plant in flower, south Wairarapa (November)

63 Mountain flax/wharariki, *Phormium cookianum,* is as common on the coasts as it is in the mountains and can be found on rocks close to the water or, more often, on steep cliffs and rocky promontories. The flowers appear from November to January, and the seeds ripen from February to March. PHORMIACEAE

64 Vittadinia, *Vittadinia australis,* is a small, low-growing, bushy plant found throughout New Zealand, growing in rocky or grassy places in coastal regions up to 900 m altitude. Flowers, about 10 mm across, occur from November till March. ASTERACEAE

65   Rangiora tree in full flower,
     Karaka Bay, Wellington
     (September)

66   Flower-head of rangiora, Kaitoke (November)

67   Close-up of rangiora flowers, Kaitoke (November)

65–67 Rangiora, *Brachyglottis repanda,* is found in coastal scrub and lowland forest, from sea-level to 800 m, throughout the North Island and in Marlborough and Nelson. Fig. 65 shows a tree in full bloom. The large, sweet-scented, much-branched panicles of flowers (fig. 66) occur from August till November; the panicles are terminal on the branches and made up of many single florets (fig. 67).

ASTERACEAE

68   Pinatoro branch with flowers,
     Cass (November)

69   Pinatoro plant in flower, Cass (November)

68–69   Pinatoro/wharengarara/common New Zealand daphne, *Pimelia prostrata,* is a prostrate or erect shrub (fig. 69) found throughout the country in rocky crags from sea-level to 1,600 m in the mountains. The sweet-scented flowers (fig. 68), about 6 mm across, occur from October till March. The drupes are either white or pinkish.

THYMELAEACEAE

71 North Cape hibiscus flower, Ahipara Bay (February)

71 North Cape hibiscus, *Hibiscus diversifolius*, is found in sandy situations from North Cape to about Ahipara Bay and Cape Brett. The perennial plant reaches to 2 m high and the flowers, up to 8 cm across, occur from November to March.

MALVACEAE

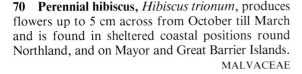

70 Perennial hibiscus flowers, Hicks Bay (October)

70 Perennial hibiscus, *Hibiscus trionum*, produces flowers up to 5 cm across from October till March and is found in sheltered coastal positions round Northland, and on Mayor and Great Barrier Islands.

MALVACEAE

72 Coastal cutty grass, *Mariscus ustulatus*, is a sedge found in damp situations in most coastal regions. It flowers during November, with the seed-heads ripening through December and January.

CYPERACEAE

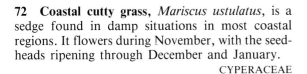

72 Coastal cutty grass with seed-heads, south Wairarapa (January)

73   New Zealand iris plant in flower, Lake Pounui
(November)

74   Seeds of New Zealand iris, Lake Pounui (May)

73–74   New Zealand iris, *Libertia ixioides*, is found
everywhere in both coastal and inland regions and
along lowland scrub and forest margins. Flowers
(fig. 73) appear during October and November, and
the berries (fig. 74) are ripe from January through
to December.   IRIDACEAE

76   Patotara with berries, Cupola Basin (April)

75–76   Patotara, *Leucopogon fraseri*, is a low,
shrubby plant found on coastal dunes, in dry, rocky
places, dry riverbeds and lowland or subalpine grasslands. Flowers (fig. 75) occur abundantly from
September to December, and the berries (fig. 76)
ripen during February and March.

EPACRIDACEAE

75   Patotara flowers, Sugarloaf, Cass, 1,000 m
(November)

77 Kopata, *Geum urbanum* var. *strictum*, grows in open country from sea-level to 900 m altitude. The flowers, 2 cm across, occur from November to January and the fruit from February to March.

ROSACEAE

77   Kopata flower and seed, Hinakura (February)

78 Giant-flowered broom, *Carmichaelia williamsii*, is a much-branched shrub that bears 2.5 cm long flowers during February and March. It is found along the Bay of Plenty coast to East Cape and on the Poor Knights, Little Barrier and Aldermen Islands.   FABACEAE

78   Flowers of giant-flowered broom, East Cape (March)

79   Close-up of some flowers of pink tree broom, Woodside Gorge (December)

80   Pink tree broom in full flower at the Woodside Gorge (December)

79–80 Pink tree broom, *Notospartium glabrescens*, when in flower (fig. 80), is one of New Zealand's most spectacular plants. Found only in Marlborough, it grows as a shrub or a small tree, up to 10 m high, on river terraces and in rocky situations from sea-level to 1,200 m. The slender, pendulous branchlets are compressed, and the flowers (fig. 79) occur as open racemes, up to 5 cm long, during December and January.   FABACEAE

81   Kaka beak flowers, Lake
     Waikaremoana (November)

82   White flowers of kaka beak,
     var. *albus*, Hinakura (October)

83   Flowers of a horticultural
     hybrid kaka beak, Hinakura
     (October)

81–83   Kaka beak/red kowhai/kowhai ngutu-kaka,
Clianthus puniceus, is a spreading shrub found wild
in the inlets of the Bay of Islands, on the coast near
Thames and at Lake Waikaremoana. The vivid red
to pale pink flowers, each about 8 cm long (fig. 81),
occur in abundance from October to December. A
white variety, *Clianthus puniceus* var. *albus* (fig. 82),
and hybrid forms (fig. 83) are also known.

FABACEAE

84   Wild lobelia, *Lobelia anceps*, is a low, herbaceous plant found in coastal and lowland areas
over most of New Zealand. Flowers, 6 mm long,
arise from November to March, blue at first but
quickly fading to pale blue or white.

LOBELIACEAE

84   Wild lobelia plant in flower, Lake Pounui
     (February)

85 Taurepo, yellow-flowered form, Hinakura (November)

86 Taurepo, red-flowered form, Hinakura (November)

85–86 Taurepo/waiu-atua/New Zealand gloxinia, *Rhabdothamnus solandri*, is a small shrub found in dry, shady, shingly gullies, along forest margins and streamsides throughout the North Island. Flowers, commonly red (fig. 86) but also yellow (fig. 85), are 18–25 mm long and occur almost all the year round.
GESNERIACEAE

87 Leafless sedge/wiwi, *Scirpus nodosus*, occurs in sandy, swampy areas on the coast and inland and around seepages on coastal hillsides.
CYPERACEAE

87 Leafless sedge with flowers, Lake Pounui (February)

88 Kaikoura rock daisy, *Pachystegia insignis*, forms a low, stiff shrub found on coastal cliffs along the Kaikoura coast and in rocky places in river valleys near the coast throughout Marlborough. The flowers, up to 7.5 cm across, appear from December to February.
ASTERACEAE

88 Kaikoura rock daisy plant in flower, Kaikoura coast (December)

89–90 Whau/cork tree, *Entelea arborescens*, is a shrub or a small, spreading tree with very large, soft, heart-shaped leaves, 15–25 cm long and 15–20 cm wide. The flowers, 18–25 mm across, arise as erect, flat or drooping clusters in the leaf axils from September to December (fig. 89) and the seeds form in spiney fruit capsules (fig. 90) from November to January. Whau is the only New Zealand tree to produce fruits with spines and these can be up to 2.5 cm long. The wood rivals balsawood as one of the lightest woods in the world. Whau occurs from North Cape south to Nelson at the bases of western coastal cliffs and on the edges of lowland forests; it is common around the Mokau River mouth.

TILIACEAE

90 Ripening seeds of whau, Hukutaia Domain (January)

89 The flowers of whau, Mokau River estuary (November)

91 Limestone harebell, *Wahlenbergia matthewsii*, is a tall herb, up to 30 cm high, that grows only on limestone rocky places between the Waima and Clarence Rivers. The flowers, 2–3 cm across, are distinctly blue when first opened but quickly fade to white. CAMPANULACEAE

91 Limestone harebell, near Woodside Gorge (December)

92   Tawapou, berries and leaves, Kaitaia (April)

92-94   Tawapou, *Planchonella novo-zelandica,* is
a small tree found along the east coast of the North
Island from North Cape to Tolaga Bay, including
the adjacent offshore islands, and to the Manukau
Harbour in the west. The axillary flowers (fig. 93)
occur in January and are very small, being 3-6 mm
across, but the berries are large, up to 2.5 cm long.
The red, yellow and orange berries (fig. 92) colour
up from March through to May.   SAPOTACEAE

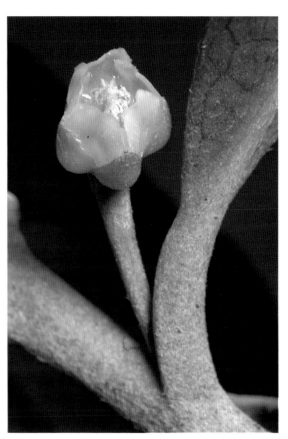

93   A flower of tawapou, Piha (January)

94   Spray of tawapou showing axillary
flowers and a green berry from last
year's flowering, Piha (January)

95   Maori jasmine in flower, Barton's Bush, Upper
Hutt (November)

95–98   Maori jasmine/kaiku/kaiwhiria, *Parsonsia
heterophylla,* found throughout New Zealand in
coastal scrub and forest and along lowland forest
margins, is a stronger, tougher vine (fig. 95) than
akakiore (opposite), producing large, white (fig. 96)
or yellow (fig. 97), peculiarly scented flowers, 8 mm
across, from September to March, followed by seed-
pods (fig. 98), 15 cm long, from February onwards.
APOCYNACEAE

96   Close-up of flowers of Maori jasmine, Stokes
Valley (October)

97   Yellow-flowered Maori jasmine, Hinakura
(November)

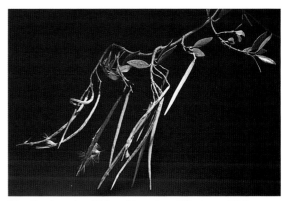

98   Seed-pods of Maori jasmine, Hinakura (February)

99–100 Akakiore/small Maori jasmine/pink jasmine, *Parsonsia capsularis*, is a slender climber (fig. 99) with small, pink, cream or white, fragrant flowers about 4 mm long (fig. 100). Found in coastal and lowland forests, it flowers from September till February. APOCYNACEAE

99 Akakiore with pinkish-coloured flowers, Hinakura (November)

100 Close-up of akakiore flowers, Hinakura (November)

101 Lowland lycopodium with strobili, Central Plateau (February)

101 Lowland lycopodium, *Lycopodium scariosum*, is a creeping plant found in lowland and subalpine forest or open scrub from the Bay of Plenty southwards, also occurring in Victoria and Tasmania. The strobili arise from the tips of ascending branches and are 2.5–5 cm long. LYCOPODIACEAE

102 Traill's daisy/tupare, *Olearia traillii*, is a handsome hybrid shrub or a small tree, 3–4 m high, the result of a cross between *O. angustifolia* and *O. colensoi*. The plant grows around the coasts of Stewart Island, flowering from November to January, and the rays of the flower-head may be violet or white. ASTERACEAE

102 Traill's daisy flower with leaves, Stewart Island (November)

103–106   Karo/turpentine tree, *Pittosporum crassifolium*, forms a shrub or a small tree, up to 9 m high, with thick, tough, leathery and wavy leaves, 5–10 cm long and up to 2.5 cm wide. It is found naturally along coastal forest margins and streamsides from North Cape to Poverty Bay and is extensively cultivated in gardens, often as a hedge plant. The scented flowers (figs 103–104) are 12 mm across and occur from September to December, with the seed capsules, 2–3 cm across (fig. 105) ripening during the following August through December. Cultivars with variegated leaves have been produced (fig. 106). On a calm evening when the tree is in full flower the rich, sweet scent from the flowers fills the air.   PITTOSPORACEAE

104   Close-up of flower of variegated form of karo, Otari (September)

103   Flowers and leaves of karo, Karaka Bay, Wellington (September)

106   Horticultural variegated form of karo in flower, Otari (October)

105   Karo seed capsules opened showing ripe black seeds, Otari (May)

107   *Pittosporum huttonianum* is somewhat similar to *P. crassifolium* but a larger tree, with leaves 12 cm long by 5 cm wide. It grows only on the Coromandel Peninsula and the Great and Little Barrier Islands, and is distinguished by the very hairy branchlets and leaf petioles as shown in fig. 107.

PITTOSPORACEAE

107   Hairy branchlet and leaf petiole typical of *Pittosporum huttonianum*, Hukutaia Domain, Opotiki (April)

108 Heart-leaved kohuhu branchlet with flowers and leaves, Kaitaia (November)

110 Flower spray of *Hebe elliptica*, Wharariki Beach, Farewell Spit (January)

109 Close-up of flowers of the heart-leaved kohuhu, Otari (October)

111 Flowers, close up, of *H. elliptica*, Wharariki Beach (January)

108–109 Heart-leaved kohuhu, *Pittosporum obcordatum*, is a narrow, erect, upward-branching, small tree, 3–4 m high, found growing wild only near Kaitaia and the Wairoa River. It is now sometimes grown in parks and gardens. The leaves are 4–8 mm long and 2–7 mm wide (fig. 108) and the flowers are borne along the branches, arising as groups from the leaf axils (fig. 109). PITTOSPORACEAE

110–111 *Hebe elliptica* forms a much-branched shrub, about 2 m high, found along the west coast from about New Plymouth southwards and along the east coast of Otago. The lateral flower racemes (fig. 110) have 4–14 flowers (fig. 111), which occur from November to March. The leaves, 2–4 cm long by 6–16 mm wide, are somewhat fleshy and keeled. SCROPHULARIACEAE

112 The flowers of manatu, Trentham (October)

113 Manatu tree in full flower, Waiorongomai (October)

112–116 Manatu/lowland lacebark, *Plagianthus regius*, is a tree up to 15 m high, with soft, coarsely toothed leaves, 10–15 cm long, many of which fall in the autumn. The tree has a juvenile form with interlacing branches that form a dense bush. Small flowers, each 3–4 mm across (fig. 114), appear in great profusion (fig. 113) as dense clusters (fig. 112) in October and November, and the downy seed capsules are ripe in January and February. Manatu is found in coastal areas and marginally in lowland forest throughout New Zealand. MALVACEAE

115 Close-up of seeds of manatu, Lake Pounui (March)

114 Close-up of manatu flowers, Trentham (October)

116 Close-up of manatu flowers, Trentham (October)

117–118 Ngaio, *Myoporum laetum,* forms a rounded tree up to 8 m high, found in coastal places and lowland forests from North Cape to Dunedin and on the Kermadec and Chatham Islands. The alternate, soft, fleshy, gland-dotted leaves, 4–10 cm, long and 10–30 mm wide, are on petioles up to 3 cm long (fig. 117). Flowers (fig. 118) about 10 mm across appear in the axils of the leaves during November and December and often again during April and May. The drupes (fig. 117) ripen three months later in each case. MYOPORACEAE

117 Ngaio berries, Karaka Bay, Wellington (March)

118 Close-up of ngaio flower, Karaka Bay, Wellington (December)

119 Prostrate ngaio, leaves and flowers, Kawhia (March)

119–120 Prostrate ngaio, *Myoporum debile,* forms a low-growing shrub often covering large areas in scrub and exposed places between Raglan and Kawhia. Fig. 119 shows a spray with flowers and typical long, narrow leaves, 4–8 cm long by 5–10 mm wide, with toothed margins towards the apex. Flowers (fig. 120), about 5 mm across, occur during March and April. MYOPORACEAE

120 Close-up of prostrate ngaio flowers, Kawhia (March)

121-125 Coastal maire, *Nestegis apetala,* is a shrub
or small tree with spreading, tortuous branches and
a furrowed bark. It is found on rocky headlands
around Whangarei Heads, the Bay of Islands, the
Hen and Chickens, Fanal, Cuvier, Poor Knights,
and Great and Little Barrier Islands. The glossy,
leathery, elliptic leaves are 4.5–12.5 cm long and
15–60 mm wide (fig. 121). The flowers are without
petals and arise as racemes of up to 21 flowers from
the leaf axils or direct from branches. Male racemes
are shown in figs 122 and 125; female racemes in
figs 123 and 124. Both flowers appear from late
January into February, and the lovely drupes (fig.
121) are ripe by the following November through
December.   OLEACEAE

121   Drupes and leaves of coastal maire, Oke Bay
      (December)

122   Raceme of male flowers of
      coastal maire, Oke Bay
      (January)

123   Raceme of female flowers of
      coastal maire, Oke Bay
      (January)

124   Close-up of female flowers of coastal maire, Oke
      Bay (January)

125   Close-up of male flowers of coastal maire, Oke
      Bay (January)

126　Parapara flowers and leaves,
Warkworth (February)

127　Close-up of flower of parapara,
Warkworth (January)

126–128　Parapara, *Pisonia brunoniana,* is the sole representative in New Zealand of a group of subtropical plants that produce sticky fruits that can capture and kill quite large birds and sometimes small reptiles. Parapara is a shrub or a small tree, up to 6 m high, with large, glossy, thick leaves, 10–40 cm long by 5–15 cm wide. Many-flowered panicles (fig. 126), with each flower up to 10 mm long (fig. 127), occur in December through January, and the sticky fruits (fig. 128) turn black and become very sticky during February and March when they ripen. Small birds can be caught and held by the sticky mass until they die. The plant then feeds upon them and the seeds may germinate on the birds' remains, the seedlings later falling to the ground to grow.　　　　　　　　　　　　NYCTAGINACEAE

129　Hooker's daisy in flower,
Stewart Island (October)

128　The sticky fruits of parapara, Warkworth (February)

129　Hooker's daisy, *Celmisia hookeri,* is a tufted herb found in coastal to montane grassland or scrub of north-east Otago and Stewart Island. The leathery leaves, 20–50 cm long by 4–8 cm wide, are smooth above but have a dense, white, appressed tomentum below. Flowers, 2–5 cm across, occur from November to January.　　　　　ASTERACEAE

130 Houhere flowers and leaves,
Karaka Bay, Wellington
(March)

131 Houhi ongaonga spray with
seeds and leaves, Karaka
Bay, Wellington (March)

130 Lacebark/houhere, *Hoheria populnea,* takes
the name lacebark from its stringy, interlaced bark
characteristic of the genus *Hoheria.* It forms a small,
erect tree up to 10 m high, with large, shining,
coarsely serrated leaves, 14 cm long by 6 cm wide
(fig. 130). The flowers, 2.5 cm across (fig. 130),
occur as 5–10-flowered cymes from February
through May. A form having flowers with purple
stamens is known as var. *osbornei* and was discov-
ered on Great Barrier Island in 1910. Houhere grows
wild only between North Cape and the Waikato in
the west and Bay of Plenty in the east; it is the only
lacebark that grows wild between Kaitaia and North
Cape It has become one of the most popular native
plants in cultivation and a variegated form has been
developed.   MALVACEAE

132 Houhi ongaonga flowers and leaves, Mt
Holdsworth (March)

131–133 Houhi ongaonga, *Hoheria sextylosa,* is a
small tree, to 6 m high, with deeply incised, serrated
leaves, 10–30 cm long by 10–25 mm wide. Flowers,
2 cm across, (figs 132–133), occur in 2–5-flowered
cymes during March and April, and seeds typical of
lacebarks (fig. 131) are produced from March
through June.   MALVACEAE

133 Close-up of flower of houhi ongaonga, Karaka
Bay, Wellington (March)

134   Narrow-leaved   lacebark/houhere/ribbon-wood, *Hoheria angustifolia*, is an attractive-looking, slender tree, to 10 m high, which matures from a straggling shrub with slender, interlacing branches bearing a few scattered leaves. The coarsely toothed adult leaves are up to 7 cm long by 10 mm wide on petioles 5 mm long. Flowers, about 15 mm across, occur singly or in 2–5-flowered cymes from December to March. Found in coastal and lowland forests from Taranaki to Southland.   MALVACEAE

136   Wharangi flowers, close up, Karaka Bay, Wellington (September)

134   Narrow-leaved lacebark spray showing leaves and flowers, Peel Forest (January)

137   Wharangi seed-pods and black seeds, Karaka Bay, Wellington (September)

135–137   Wharangi/koheriki, *Melicope ternata*, is a spreading and branching shrub or a small tree, up to 6 m high, found along coastal and lowland forest margins and on sheltered, rocky places near the sea from North Cape to Nelson and Kaikoura. The leaves (fig. 135) are trifoliate, wavy, 7–10 cm long by 3–4 cm wide, on petioles 2–3 cm long. Flowers (fig. 136) arise as paired, three-branched cymes from September to November, the single flowers each 10 mm across. The black seeds (fig. 137) appear from September to October.   RUTACEAE

135   Wharangi leaves with developing flower-buds, Karaka Bay, Wellington (February)

138 Poataniwha flowers and typical leaf, Trentham (October)

139 Close-up of male flowers of poataniwha, Barton's Bush, Trentham (November)

140 Stewart Island daisy in flower (December)

138-139 Poataniwha, *Melicope simplex*, is a twiggy shrub about 3 m high found in coastal situations throughout New Zealand up to 600 m altitude. Young plants have trifoliate leaves, up to 10 mm long on petioles 2 cm long, but mature plants have simple, rhomboid-leaves, up to 2 cm long by 2 cm wide, on petioles 5 mm long (fig. 138). Flowers (female, fig. 138; male, fig. 139) occur from September through November, and the seeds, similar to those of *M. ternata* but smaller, occur from December through to April. RUTACEAE

140 Stewart Island daisy, *Celmisia rigida*, is a tufted herb with rigid, leathery leaves, 10–15 cm long by 5–7 cm wide, found on coastal cliffs of Stewart Island. Flowers, 4–5 cm across, occur during November and December. ASTERACEAE

141 White harebell plant in flower, Charleston (February)

141 White harebell, *Wahlenbergia congesta*, is a creeping herb forming dense patches to 25 cm across on lowland grassy banks in sand-dune country from Cape Foulwind to Charleston. A varietal form is found on Mt Taranaki. CAMPANULACEAE

142 Pink broom in full flower, Clarence River (December)

143 Chatham Island forget-me-not, *Myosotidium hortensia*, is a handsome perennial with large, glossy, kidney-shaped leaves, 25–40 cm wide, each with a stout petiole, which, with the leaf, can be 1 m long. Flowers are in dense cymes, 10–15 cm across, on stalks up to 60 cm high, from September to November. Occurring naturally only on the Chatham Islands, it is now a very highly prized garden plant.
BORAGINACEAE

143 Chatham Island forget-me-not plant in flower, Otari (November)

142 Pink broom, *Notospartium carmichaeliae*, grows as a shrub or a small tree, up to 10 m high, with drooping, leafless branches and compressed branchlets that become smothered with small, crowded racemes of pink-flushed and pink-veined flowers, each flower about 8 mm long, during December and January. FABACEAE

PLANTS OF LOWLAND SCRUB

Scrub is an association of shrubby and herbaceous plants and occasional small trees found in coastal regions; sometimes it is open but is more often dense and perhaps impenetrable. Scrub forms strange and peculiar associations of plants almost unique to New Zealand. It is generally encountered in gullies or covering dry, windswept hillsides, dry coastal terraces or river terraces and plateaus. It can be the forerunner of forest, providing a sheltered, nourishing seed-bed for larger trees. Many plants found in scrub also occur in forest or along forest margins, and the species associations can vary considerably, even within small areas, or be particularly locally defined. Lowland scrub can extend inland for 1–2 km from the coast and up to elevations of 700 m.

144 Flower-heads of golden tainui, from my Waikanae garden (September)

144–146 Golden tainui/kumarahou/gum-diggers' soap, *Pomaderris kumaraho,* is a rounded shrub bearing large, golden yellow corymbs of flowers (fig. 144) in profusion during September and October. It is found growing on poor soils from North Cape to Tauranga on the east coast and to Kawhia on the west. Leaves 6 cm long by 3 cm wide are soft, with depressed veins and star-like hairs beneath. A somewhat similar, more open shrub with leaves having fine tomentum beneath and the flower-heads smaller and more open is found round Warkworth to Thames and is known as *P. hamiltonii* (fig. 146).

RHAMNACEAE

145 Close-up of golden tainui flowers, Waikanae (September)

146 Spray of *Pomaderris hamiltonii* showing leaves and flower-head, Waipahihi Botanical Reserve, Taupo (September)

147–148 Wrinkled-leaved pomaderris, *Pomaderris rugosa*, is an erect shrub found growing on poor soils around Northland, the Coromandel Peninsula and Mayor Island. It is distinguished from the golden tainui by its wrinkled leaves (fig. 148) and longer flower-heads (fig. 147), and from tainui by the clothing of close, stellate hairs on the undersides of the leaves and the less-prominent leaf veins (fig. 148).

RHAMNACEAE

149 Flower-head of tainui, Kawhia (November)

147 Flower-head of wrinkled-leaved pomaderris, Coromandel Peninsula (November)

150 Spray of tainui showing prominent leaf veins below, Kawhia (November)

149–150 Tainui, *Pomaderris apetala*, is an erect, branching shrub up to 4 m high, with wrinkled, crenulate-margined, prominent-veined leaves, 7 cm long and 3 cm wide, hairy below with occasional hairs above (fig. 150). The greenish flowers (fig. 149), without petals, occur as lateral and terminal clusters on the branchlets from November into January. Tainui is an Australian shrub that grows in New Zealand naturally from Kawhia to Waitara and has become popular and widespread in gardens.

RHAMNACEAE

148 Undersurfaces of leaves of wrinkled-leaved pomaderris showing stellate hairy surface, Coromandel Peninsula (October)

151   Dwarf pomaderris/tauhinu, *Pomaderris phylic-ifolia*, is a low-growing, heath-like plant that smothers itself with corymbs of yellow flowers during October and November. The small, thick leaves, with recurved margins, are 10 mm long by 3 mm wide and densely hairy below. There are two varieties, var. *ericifolia*, with flower corymbs laterally along the branches, and var. *polifolia*, (fig. 151), which has flower corymbs both along the branches and at their tips.

RHAMNACEAE

151   Flowers of dwarf pomaderris var. *polifolia*, Whangarei Heads (October)

152   Poroporo flowers, Karaka Bay (October)

154   Spray of manuka flowers, Lake Pounui (December)

153   Poroporo berries, Karaka Bay (January)

152–153   Poroporo, *Solanum aviculare*, forms a branching shrub up to 3 m high. Along with the very similar *S. laciniatum* (fig. 272, p. 68), the flowers (fig. 152), are up to 3.5 cm across and occur from September till April. The berries (fig. 153) ripen from November onwards. It is found growing in frost-free areas from Auckland to Dunedin and on the Kermadec Islands.

SOLANACEAE

155   Close-up of manuka flower, Lake Pounui (November)

156   Manuka seeds, Woodside Gorge
(September)

158   Close-up of flower of manuka var. 'Martinii',
Otari (September)

159   Flowers of manuka var. *keatleyi*, Otari
(December)

157   Flower spray of horticultural manuka var.
'Ruby Glow', Otari (November)

154–159 Manuka, *Leptospermum scoparium*, is probably one of the best-known New Zealand plants, growing as a shrub or a small tree throughout both the North and South Islands. Manuka flowers (figs 154–155) normally are about 12 mm across but the variety *keatleyi* (fig. 159) has flowers 2 cm across. Flowers occur from September till February, in the wild usually white (fig. 155), but coloured wild forms have occurred, such as var. *keatleyi*, and from these several fine horticultural forms with single or double flowers have been developed; var. 'Ruby Glow' (fig. 157) and var. 'Martinii' (fig. 158) are good examples. The characteristic seeds of manuka are shown in fig. 156; they ripen during April and May but persist on the tree until the following year. The flowers of the related tree **kanuka,** *Leptospermum ericoides*, are borne much more densely, and the stamens spread more than they do in manuka flowers.

MYRTACEAE

160 Close-up of flowers of kanuka, Taupo (November)

160–163 Kanuka/teatree, *Leptospermum ericoides,* is a tall shrub or a small tree, up to 15 m high, which grows throughout the North and South Islands from sea-level to 900 m altitude. The acute, narrow, aromatic leaves are 12–15 mm long by 2 mm wide. The small flowers (fig. 160), about 5 mm across, clothe the branches in great profusion from September into February (figs 162–163), and the seeds, similar to those of manuka, remain on the tree until the following year. MYRTACEAE

161 A kanuka tree in full flower, Lake Rotoiti, Nelson (January)

162 Spray of kanuka flowers showing density of flowering, Taupo (January)

163 Spray of kanuka flowers from Hinakura (December)

164–165 Red-fruited karamu, *Coprosma rhamnoides*, is one of the small-leaved, twiggy coprosmas, with interlacing branches forming tight bushes (fig. 164). It is found growing in lowland and subalpine scrub and forests all over New Zealand. The drupes (fig. 165) appear in early November as small, red berries, 3–4 mm in diameter, which turn dark crimson or black as they ripen. RUBIACEAE

164 Red-fruited karamu bushes growing on windswept hillside, Wellington (February)

165 Ripe red-fruited karamu drupes, Pahaoa Valley (December)

166 Karamu drupes, Opepe Bush (March)

167 Male flowers of karamu, Karaka Bay (October)

166–168 Karamu, *Coprosma lucida*, is a small tree, 3–6 m high, found in scrub, along forest margins and in forests throughout the North, South and Stewart Islands from sea-level to 1,060 m altitude. One of the larger-leaved coprosmas, its glossy, leathery leaves, 12–19 cm long by 3–4 cm wide, are dark green above but paler below, with very prominent domatia on the lower surface. The flowers (fig. 167, male flowers; fig. 168, a spectacular group of female flowers) appear from September into October, while the drupes (fig. 166), some 12 mm long, mature during the following winter to ripen by February and March, some 14 months after the flowering.

RUBIACEAE

168 A magnificent display of karamu female flowers, Karaka Bay (October)

169-171 Kakaramu, *Coprosma robusta*, is probably the most widespread coprosma species found throughout New Zealand in scrub and forests. Kakaramu grows as a shrub or a small, spreading tree up to 6 m high, with stout, hairless branches and large, leathery leaves (fig. 169), 12 cm long and 5 cm wide. Flowers (fig. 169), occur from September till November. The brilliant red drupes are ripe by March and smother the tree with colour well into the winter (figs 170-171).   RUBIACEAE

171   Kakaramu drupes, close up, Waipunga Gorge (April)

169   Male flowers of kakaramu, Taupo (October)

172-173 Stinkwood/hupiro, *Coprosma foetidissima*, is a shrub or small tree (fig. 173) found growing in scrub, along forest margins and in forests from sea-level to 1,360 m throughout New Zealand. The quite large, solitary flowers, about 10 mm long, occur during September and October, and the drupes, 7-10 mm long, arise singly along the branches and ripen during May and June (fig. 172). When crushed, or even brushed against, the plant emits a vile odour like that of many rotten eggs. With its fresh, rather fleshy looking foliage and brilliant red, single drupes, stinkwood, however, looks quite striking.   RUBIACEAE

170   Dense clusters of drupes clothe a branch of kakaramu, Taupo (March)

172   Stinkwood drupe and leaves close up, Mt Holdsworth (June)

173 Stinkwood tree on forest margin,
Mt Ruapehu (January)

174 Hairy coprosma leaves
showing upper sides, Ure
River Gorge (January)

174–176 Hairy coprosma, *Coprosma crassifolia*, grows as a shrub or a small tree, with spreading, stiff, interlacing branches, hairy when young. It is found in scrub, tussocklands, river terraces, rocky places and forests from sea-level to 400 m southwards from Mangonui through both islands. The leaves (figs 174, upper surfaces; 175, lower surfaces showing hairy margins and tomentum) are small, thick and rounded, with the petioles and leaf margins very hairy. The white to cream-coloured drupes (fig. 176), 5–6 mm long, are ripe in March and April.

RUBIACEAE

175 Hairy coprosma leaves
showing lower surfaces, Ure
River Gorge (January)

176 Twigs of hairy coprosma with leaves and drupes,
Ure River Gorge (March)

177 ***Coprosma macrocarpa*** grows as a shrub or a small tree, 5–10 m high, and is found on the Three Kings Islands, and from North Cape to Kaipara Harbour. The wavy, leathery leaves are 9–13 cm long by 4–8 cm wide and have a prominent mid-vein. The drupes, orange-red when ripe, are 10–25 mm long and ripen during December and January.

RUBIACEAE

177 *Coprosma macrocarpa*, branch showing leaves and immature drupes, Warkworth (December)

178 Maruru/hairy buttercup, *Ranunculus hirtus*, is a tall, very hairy-stemmed and hairy-leaved buttercup, up to 60 cm high, found in scrub, often in rocky places, and in forest throughout New Zealand. The flowers are 15 mm across and occur from September till February.

RANUNCULACEAE

178 Maruru flowers, Mt Egmont (December)

180 Flowers and flower-buds of *Cassytha paniculata*, Cape Reinga (January)

179–180 ***Cassytha paniculata*** grows parasitically on herbs and low shrubs, forming a tangled mass of stems that sprawls over everything nearby (fig. 179). It is found in scrubland from North Cape south to Ahipara and Mangonui. The minute, sessile flowers arise in succulent perianth tubes (fig. 180), and these and the fruits occur from November through till April.

LAURACEAE

179 *Cassytha paniculata* plant sprawling over low shrubs, Cape Reinga (January)

181 Tutu flowers, Taupo (January)

181–185 **Tutu,** *Coriaria arborea*, is a shrub or small tree, up to 8 m high, that is very poisonous to humans and other animals. In spring the sap and in autumn the seeds contain a poisonous glucoside 'tutin', which makes tutu New Zealand's most poisonous native plant. One milligram of tutin is sufficient to severely upset a healthy person. Tutu occurs all over New Zealand and the Chatham Islands, from sea-level to 1,060 m, in scrublands, along forest margins, in gullies and on river terraces and is recorded as being one of the first plants to appear after a forest fire. The stems are four-sided and the opposite leaves are 5–10 cm long by 4–5 cm wide (fig. 185). The flowers (fig. 181) occur as pendulous racemes, up to 30 cm long, from September till February (fig. 182 shows a male flower; fig. 183 a female flower), and the berries (fig. 184) ripen from November through April. CORIARIACEAE

182 Close-up of male tutu flower, Taupo (October)

183 Close-up of female tutu flower, Karaka Bay (November)

184 The ripe poisonous berries of tutu, Hanmer (January)

185 A tutu leaf, Waipahihi Botanical Reserve (February)

186   Spray of korokio flowers, Waikanae (November)

187   Close-up of korokio flower, Waikanae (November)

188   Leaves and yellow berries of korokio, Waikanae (April)

189   Red berries of korokio, close up, Waikanae (April)

186–190   Korokio, *Corokia cotoneaster*, is a much-branched shrub, up to 3 m high, with stiff, often interlacing branches, found in scrub and on dry river flats and rocky places throughout New Zealand and on the Three Kings Islands. The leaves (fig. 186) are small, 2–15 mm long by 2–10 mm wide, on flattened petioles. Flowers (figs 186–187) appear in profusion from September through December, and the drupes, which may be either yellow or red (figs 188–189), ripen from January through May. A related plant, *C. buddleioides*, which is found in the northern parts of the North Island and which has large leaves similar to those of *C. macrocarpa* (page 67), sometimes hybridises with *C. cotoneaster* to produce a hybrid form known as var. *cheesemanii* (fig. 190).
CORNACEAE

190   Korokio var. *cheesemanii* in full berry, Otari (August)

191 Bracken fern, *Pteridium aquilinum* var. *excelsum*, can grow luxuriantly to a height of 4 m, often covering large areas as pure stands in scrub, and along forest margins throughout New Zealand. During December through January the young uncurling fronds can be quite a spectacular sight.
PTERIDACEAE

191   Bracken fern, Taupo (January)

192 The hairy nertera, *Nertera dichondraefolia*, forms patches 40 cm across in dry scrub, lowland forests, and sometimes in grasslands from Coromandel Ranges south to Stewart Island. Recognised by its hairy stems and leaves, it flowers from October till February and carries these berries from December to May.
RUBIACEAE

192   The hairy nertera with drupes, Otira Gorge (April)

193   Powhiwhi flowers, Akatarawa (December)

193–194 Powhiwhi/New Zealand convolvulus, *Calystegia tuguriorum*, and **pohue/rauparaha,** *Calystegia sepium*, are both creeping plants, with powhiwhi usually creeping higher than pohue. Powhiwhi has its leaves on petioles 4 cm long and flowers from November to February, while pohue has leaf petioles 10 cm long and flowers from October to March.
CONVOLVULACEAE

194   Pohue flowers, Taupo (February)

195 Common hairy nertera, *Nertera setulosa,* is a very hairy to setose plant, forming patches 30 cm across in scrub, exposed open places and along forest margins from North Cape to Stewart Island. It flowers from November to February, with red drupes from January through May. RUBIACEAE

199 Native broom, *Carmichaelia aligera,* is found commonly in scrub and along forest margins throughout the northern half of the North Island. Fig. 199 shows stems bearing some ripened seed-pods, which appear during March and April but often remain attached to the plant until November. FABACEAE

195 Common hairy nertera plant with flower, near Opotiki (November)

196 Ciliated nertera plant with flowers and drupes, Mt Egmont (November)

196 Ciliated nertera, *Nertera ciliata,* forms patches 20 cm across in scrub and lowland to montane forests from Arthur's Pass southwards. The heart-shaped leaves have occasional cilia-like hairs. Flowers and drupes occur from October till March. RUBIACEAE

197 Tutukiwi flowers, Akatarawa (December)

197 Tutukiwi/hooded orchid/elf's hood, *Pterostylis banksii,* is a ground orchid found throughout New Zealand in shady places in lowland and subalpine scrub and forests. This orchid flowers from September through December, with the flowers, 5–7.5 cm long including the tails, on stems 30–45 cm high. ORCHIDACEAE

198 Grassy hooded orchid, *Pterostylis graminea,* is a ground orchid, usually about 20 cm high, with grass-like leaves. It flowers from September to January and is found in similar situations to tutukiwi. ORCHIDACEAE

198 Grassy hooded orchids in flower, Hinakura (December)

201   Thick-leaved tree daisy, *Olearia pachyphylla*, has thick, leathery, smooth leaves, 7–12.5 cm long by 5–6.5 cm wide. This olearia is found naturally from the Bay of Plenty southwards in lowland to montane scrub and forests. Flowers occur in large corymbs from October through February, and the fruits ripen from November till March.

ASTERACEAE

199   Native broom seeds in seed-
pods, Whangarei (September)

201   Spray of thick-leaved tree daisy with flowers, Hukutaia Domain (January)

200   Tamingi spray in flower,
Whangarei (October)

202   Koromiko taranga, spray with flowers and leaves, Rimutaka Hill (February)

200   Tamingi, *Epacris pauciflora*, forms an upright, slender shrub, up to 2 m in height, with flowers, 8 mm across, tightly packed around the stems from October through April. It is found in lowland scrub, and boggy places from North Cape south to Marlborough.   EPACRIDACEAE

202   Koromiko taranga, *Hebe parviflora*, showing flowers and leaves. It is a shrub or a small tree, much-branched with twiggy branches and narrow leaves 2.5–6.5 cm long. Flowers occur on stalks 5–10 cm long from December through March. The plant is found from Auckland south throughout both islands from sea-level to 600 m altitude.

SCROPHULARIACEAE

203–206 Mingimingi, *Leucopogon fasciculata,* is an open-branched shrub or tree to 6 m high, found from the Three Kings Islands to Canterbury in lowland scrub or forest, and in rocky places from sea-level up to 1,150 m. The narrow, lanceolate leaves, 12–25 mm long by 2–4 mm wide, are spreading and sharp pointed (fig. 206). Flowers occur as drooping racemes (fig. 203) or spikes, each with 6–12 small flowers from August through to December (fig. 204). The fruits (fig. 205) ripen from September to April or May.   EPACRIDACEAE

205 Mingimingi (*L. fasciculata*) with ripening berries, Taupo (February)

203 Branchlet of mingimingi (*Leucopogon fasciculata*) with terminal raceme of flowers, Taupo (October)

206 The leaves of mingimingi, (*L. fasciculata*), showing the veins and the hairy stem, Taupo (February)

207–211 Mingimingi, *Cyathodes juniperina,* is an open-branching shrub to 5 m high, with narrow pungent leaves, 6–20 mm long by about 1 mm wide, with pale, white-striped lower surfaces (fig. 209). The flowers, unlike those of *L. fasciculata,* occur singly along the branches (figs 207–208) from August till December, and the fruits ripen from October through March, occurring in varying shades from white through pink to red (figs 210–211), and persisting on the branches till July. *C. juniperina* is found in similar situations to *L. fasciculata* but also occurs further south to Stewart Island.

EPACRIDACEAE

204 Close-up of mingimingi (*L. fasciculata*) flowers showing the intricate, beautiful structural detail, Taupo (October)

209 Undersides of leaves of
C. juniperina, showing
typical white stripes of this
species, Volcanic Plateau
(October)

207 Mingimingi *(C. juniperina)*, flowers
around tip of branchlet, Volcanic
Plateau (October)

210 Sprays of mingimingi (*C. juniperina*) with
berries, Volcanic Plateau (March)

208 Mingimingi (*C. juniperina*), showing flowers
along the branchlet, Volcanic Plateau (October)

211 Mingimingi (*C. juniperina*) with white berries,
Lake Pounui (February)

212 *Hebe ligustrifolia* is a rather openly branched shrub, found from Dargaville northwards in scrub and along forest margins. The leaves are up to 5 cm long, and the flowers, occurring during December and January, are about 10 cm long.

SCROPHULARIACEAE

212 Spray of leaves and flowers of *Hebe ligustrifolia*, near Waipoua (December)

214 Chatham Island tree daisy, *Olearia chathamica*, has finely serrate leaves, 2.5–8 cm long, and flowers 3–4.5 cm across, produced in abundance from October till February. It is found in both dry and damp scrub on the Chatham Islands.

ASTERACEAE

214 Flowers of Chatham Island tree daisy, Otari (November)

215–216 Akiraho/yellow akeake, *Olearia paniculata*, is a small tree up to 6 m high, readily recognised by its crinkly leaves (fig. 216), 3–10 cm long by 2–4 cm wide, and its sweet-scented flowers (figs 215–216), which occur during March and April. It is found in scrub from East Cape to Greymouth and Oamaru. ASTERACEAE

213 Koromiko/koromuka/willow koromiko, *Hebe stricta,* is a slender, much-branched shrub, up to 4 m high, with grey bark and willow-like, sessile leaves, 10–15 cm long. The flowers on stalks up to 20 cm long occur from December till March. The shrub is found throughout the North Island from sea-level to 1,100 m. A variety *atkinsonii* is found in coastal Marlborough. A similar species, *H. salicifolia*, is found throughout the South Island.

SCROPHULARIACEAE

213 Koromiko flowers, Rimutaka Hill (February)

215 Close-up of flowers of akiraho, Karaka Bay (March)

217-218 Twiggy tree daisy, *Olearia virgata,* is a twiggy shrub or a small tree with four-angled branchlets (fig. 218) and small leaves arising in widely separated, opposite fascicles of two to four leaves. Flowers (fig. 217), about 9 mm across, occur in abundance from October till January. *O. virgata* is found in damp or boggy ground in scrub from Thames south to Stewart Island.

ASTERACEAE

218 Branchlet of twiggy tree daisy var. *implicata* from the Port Hills, Christchurch (January)

217 Twiggy tree daisy flowers, close up, Otari (November)

219-220 Teteaweka, *Olearia angustifolia,* is a shrub or a small tree, up to 6 m high, found in coastal scrub along the coasts of Southland, Foveaux Strait and the headlands of Stewart Island. The thick, narrow-lanceolate leaves, up to 15 cm long and 2 cm wide, have crenate dentate margins and a thick, soft, white tomentum below (fig. 219). The large, scented flower-heads, 3.5–5 cm across, occur from October through January (figs 219–220). ASTERACEAE

219 Teteaweka plant with flowers grown at Happy Valley, Wellington, by Mrs Natusch (November)

216 Akiraho flowers, Karaka Bay (March)

220 Close-up of flower from plant in fig. 219.

221   Spray of fragrant tree
       daisy with flowers, Otari
       (November)

222   Close-up of flowers of
       fragrant tree daisy
       showing also the ridged,
       four-angled branchlet,
       Otari (November)

221–222   Fragrant tree daisy, *Olearia fragrantissima*,
forms an erect, much-branched shrub or small tree,
up to 5 m high, with zig-zagging branches bearing
variable, more or less elliptic leaves 7–30 mm long
by 5–10 mm wide (fig. 221). The delightfully fragrant
flowers (fig. 222) occur in dense clusters about 2 cm
across from October through February. *O. fragrantissima* is found in east coast lowland scrub from
Banks Peninsula southwards.   ASTERACEAE

223–224   Deciduous tree daisy, *Olearia hectori*,
found in scrub from sea-level to 900 m around
Taihape and from the Clarence River to Southland,
is a shrub or a small tree up to 5 m high and has
slender, rounded, slightly grooved, smooth branchlets. The narrow to broad, obovate leaves, 2.5 cm
long by about 10–20 mm wide, are in fascicles of 2–4
on short branchlets (fig. 224) Flowers occur from
October through December as drooping fascicles on
silky pedicels, each with 2–5 heads about 5 mm
across (fig. 223).   ASTERACEAE

223   Flowers, close up, of deciduous
       tree daisy, Otari (November)

224   Leaf spray of deciduous
       tree daisy, Otari (November)

225–226 Akepiro/tanguru, *Olearia furfuracea,* occurs as a shrub or small tree 5 m high from sea-level to 900 m along forest margins, streamsides and in scrub from North Cape to the Southern Ruahine mountains. The branchlets are angled, grooved and pubescent, with the thick, leathery, wavy-margined leaves, 7–13 cm long by 5–6.5 cm wide, on petioles 2.5 cm long, having a buff-coloured, satiny tomentum below (fig. 225). Flowers are sweet scented and occur as large, flat corymbs from November to February (fig. 226).   ASTERACEAE

226   Flowers of akepiro, Taupo
(December)

225   Leaf spray of akepiro, Taupo
(December)

227–228 Common tree daisy, *Olearia arborescens,* is a shrub 4 m high with angled branchlets and thin, leathery leaves, 2–6 cm long by 2–4 cm wide, with a thin, satiny tomentum below and a 2 cm long petiole (fig. 228). Flowers (fig. 227) are in large, flat corymbs, 25 cm across, produced in profusion from October through January. This is one of the most common and beautiful of the *Olearia* species and is found from East Cape to Stewart Island.

ASTERACEAE

228   Leaves of common tree
daisy, Whangamoa Saddle
(November)

227   Flower-head of common tree daisy, Whangamoa
Saddle (November)

229-230 Akeake, *Olearia avicenniaefolia,* is a shrub or small tree 6 m high, with narrow, elliptic, slightly wavy, thick leaves, 5–10 cm long by 3–5 cm wide, shining above but with a soft, white or fawn-coloured tomentum below (fig. 229). The flowers are in large flat corymbs on long stalks and occur from January through April (fig. 230). The shrub is found from sea-level to 900 m in scrub throughout the South Island and Stewart Island.

ASTERACEAE

230 Flowers of akeake, Waipunga Gorge (April)

229 Leaves of akeake with flower-buds, Lewis Pass (December)

231-233 Heketara/forest tree daisy, *Olearia rani,* is a shrub or a small tree 7 m high found in scrub and lowland forest, along forest margins, river and stream banks, in forest clearings and in second-growth forest from North Cape to Nelson and Marlborough (fig. 231, in Otaki Gorge). It is one of the most beautiful of our tree daisies. The characteristic and unmistakable leathery, dentate leaves, 5–15 cm long by 5 cm wide, (fig. 232), on petioles 4 cm long, are, like the branchlets and petioles, covered below with a dense whitish or brownish, soft tomentum. The flowers (fig. 233) appear from August through November as large sprays up to 18 cm across, the individual ray florets 10–20 mm across.

ASTERACEAE

231 Heketara flowering in scrub, Otaki Gorge (October)

232 Heketara leaves, Stokes Valley (October)

234 Scented tree daisy *Olearia odorata*, is an erect, spreading shrub, up to 4 m high, with divaricating branches and rounded leaves, 10–25 mm long by 6-15 mm wide, which arise in opposite fascicles. Small, strongly sweet-scented flowers, 8–12 mm across, occur in fascicles along the branches from December to February, and the shrub is found from Kaikoura southwards in montane and subalpine scrublands.   ASTERACEAE

234   Branches of scented tree daisy with flowers and leaves, Otari (February)

233   Heketara flowers, close up, Kaitoke (November)

235–237 Inanga/grass tree, *Dracophyllum longifolium*, is the most widespread grass tree, being found in lowland and subalpine scrub and forest from East Cape southwards through the North, South and Stewart Islands, the Auckland Islands and the Chatham Islands. It grows as a slender, erect tree, 12 m high, with spreading branchlets, which have the stiff leaves crowded towards their tips. The leaves are 10–25 cm long by 3–5 mm wide, tapering to a long, acuminate tip (fig. 236). The shape and form of the basal leaf sheath (fig. 237) is very useful in identifying species of *Dracophyllum*. Flowers in 6–15-flowered racemes (Fig. 235) appear terminally on the branchlets during November and December.
                    EPACRIDACEAE

235   Inanga flowers, Mt Dobson (December)

236   Inanga leaves, Otari (April)

237   Inanga leaves, showing basal leaf sheaths, Otari (April)

238-240　*Dracophyllum lessonianum* is a small tree, 10 m high, found in scrub from North Cape to Kaitaia often among manuka. Fig. 238 shows a tree, fig. 239 the leaves, 6–10 cm long by 1–1.5 mm wide, and fig. 240 shows the flower racemes, each about 3 cm long, which arise singly or in clusters terminally on short, lateral branches.　　　EPACRIDACEAE

240　Flowers, close up, of *D. lessonianum*, Kaitaia (April)

238　A small grass-tree, *Dracophyllum lessonianum*, Kaitaia (April)

241-242　*Dracophyllum sinclairii* is a slender tree, 3–6 m high, with close-set leaves, 3.5–12.5 cm long by 4–8 mm wide, which have finely serrulate margins and long, acuminate apices (figs 241–242). Flowers in 4–8-flowered racemes arise terminally and subterminally (fig. 241) on lateral branches during April and May. The tree is found from North Cape south to Kawhia.　　　EPACRIDACEAE

239　Leaves and flower-buds of *D. lessonianum*, Kaitaia (April)

241　Tip of branchlet of *Dracophyllum sinclairii* with flowers, Kaitaia (May)

243-245 *Dracophyllum viride* is a small tree, up to 5 m high, with slender, ascending, leafy branches bearing long (juvenile) and short (adult) leaves. It is found in scrub only between North Cape and Kaitaia. Adult leaves, 5-7 cm long by 5-6.5 mm wide, tend to spread and have a distinct ciliated shoulder at the top of the sheath (fig. 244). Juvenile leaves occur on both young and old plants and are 2-3 times longer than adult leaves (fig. 243). Flower racemes, each with 5-6 flowers, arise just below leaf tufts on the branchlets (fig. 245) during March and April.   EPACRIDACEAE

244   The ciliated shoulder on the leaf of *D. viride*, Herekino Saddle, Kaitaia (April)

243   Juvenile and adult foliage of the grass-tree, *Dracophyllum viride*, Herekino Saddle, Kaitaia (April)

242   The shoulder in the leaf of *D. sinclairii*, Kaitaia (April)

245   Growing apex of grass-tree, *D. viride*, showing adult leaves below with flowers and juvenile leaves above, Herekino Saddle, Kaitaia (April)

246–247 **Lesser caladenia,** *Caladenia carnea* var. *minor*, is a hairy ground orchid, 5–16 cm high, usually with a single, narrow leaf, found throughout the North Island, sometimes concentrated in large numbers in dry places under open scrub canopy and on dry, open clay hills. The flowers, 1–2 cm across, occur from October through January, varying in colour from bluish white (fig. 246) through pinks (fig. 247) to purple.   ORCHIDACEAE

248 **Giant sedge,** *Gahnia xanthocarpa*, has densely tufted stems reaching to 4 m in height and leaves 2 cm wide, and is the largest sedge found in New Zealand. It forms large clumps on the edges of bogs, in scrub, along forest margins and inside forests, especially in Northland, but also discontinuously throughout both North and South Islands. The flower panicles, 60 cm to 1.5 m high, appear from December through to March.   CYPERACEAE

246   The lesser caladenia, greenish white form, Hinakura (December)

247   The lesser caladenia, pink form, Hinakura (December)

248   Giant sedge plants in flower, Lake Pounui (December)

249   Flowers of *Teucridium parviflorum*, Hinakura (December)

249   *Teucridium parviflorum* is a much-branched, twiggy shrub, sometimes forming thickets up to 1.5 m high or sprawling over rocks in lowland scrub and forest throughout New Zealand. The branchlets are four-sided and hairy when young; the small leaves, 4–15 mm long, are on petioles almost as long as themselves. Solitary axillary flowers, 8 mm across, occur from October through January.

VERBENACEAE

250 Papataniwhaniwha, *Lagenophora pumila*, is found along lowland forest margins and in scrub or grasslands from Rotorua southwards. This is one of five species of *Lagenophora* found in New Zealand and has leaves 2–6 cm long.   ASTERACEAE

250   Papataniwhaniwha plant in flower, Waihohonu Stream, Tongariro National Park (January)

252   Spray of ongaonga with female flowers and showing stinging hairs, Hinakura (November)

251–254 Ongaonga/tree nettle, *Urtica ferox*, is a soft-wooded shrub or small tree, up to 3 m high, with many intertwining branches. The leaves, branchlets and branches all bear stout stinging hairs that can inflict a painful sting, and, if stung many times, a person can lose co-ordination of muscle movement for upwards of three days; deaths from the stings are recorded. Ongaonga is found in scrub and along forest margins, forming thickets from sea-level to 600 m. Leaves are 8–12 cm long by 3–5 cm wide, and fig. 253 shows a leaf underside with stinging hairs and female flowers. Branchlets, leaves and female flowers are shown in fig. 252; male flowers are shown close up in fig. 251 and females flowers close up in fig. 254. Flowers are in spikes about 8 cm long and occur from November through March.

URTICACEAE

253   Underside of ongaonga leaf, showing stinging hairs and female flowers, Hinakura (November)

251   Male flowers of ongaonga, Hinakura (November)

254   Close-up view of ongaonga female flowers, Hinakura (November)

255–256 Rohutu, *Neomyrtus pedunculata*, is a shrub or small tree up to 6 m high, with four-sided branches that are hairy when young. It occurs in lowland scrub and along forest margins all over New Zealand from sea-level to 1,050 m. The leaves are of variable shape, obovate-oblong and 15–20 mm long by 10–15 mm wide or obovate and 6–15 mm long by 4–10 mm wide, all dotted with glands, thick and leathery with rolled or thickened margins. Solitary flowers, 5 mm across (fig. 256), arise from leaf axils from December through to April, and the berries (fig. 255) are ripe from February through May.

MYRTACEAE

255 Branch of rohutu with berries, Ponatahi (April)

256 Flowers and leaf of rohutu, Makarora Valley (January)

257 Flowers and leaves of *Pittosporum lineare*, Kaimanawa Mountains (October)

257 *Pittosporum lineare* is a much-branched, divaricating shrub about 3 m high found in lowland scrub and along forest margins throughout the North Island, Nelson and Marlborough. The fascicled leaves, 10–15 mm long by 1–3 mm wide, are paler below. The small black flowers, about 5 mm across, arise singly and mostly terminally on branchlets during October through January.

PITTOSPORACEAE

258–259 *Neopanax anomalum* is a shrub, 3 m high, with densely divaricating branches. The branchlets have fine bristle-like setae and small, rounded leaves, 10–20 mm long by 10–15 mm wide, on petioles 5 mm long. Minute flowers (fig. 258) in 2–10-flowered umbels occur from November to February, with berries ripe during March and April (fig. 259).

MYRTACEAE

258 Flowers of *Neopanax anomalum*, Rahu Saddle (December)

260-263 Horopito/pepper tree, *Pseudowintera colorata*, is commonly a shrub but sometimes a small tree, up to 10 m high, noted for its reddish coloured, aromatic leaves, 2–8 cm long by 10–30 mm wide (fig. 260), with pungent taste. It is common in lowland scrub and forests as well as in alpine regions up to 1,200 m throughout New Zealand. Aromatic flowers (figs 261–262) occur singly or in fascicles along the branches, each on a pedicel 10 mm long, from October to March, and the dark reddish black fruits (fig. 263) are ripe from December to June.

WINTERACEAE

260   Foliage of horopito, Hihitahi State Forest (April)

261   Branchlets of horopito with flowers and showing typical blue-coloured leaf undersurfaces, Opepe Bush (October)

262   Close-up of horopito flowers, Waipunga Gorge (October)

259   *N. anomalum* showing branchlets with leaves and berries, Rahu Saddle (April)

263   Spray of horopito with berries, Waipunga Gorge (April)

264   Northern cassinia plant in
flower, North Cape
(January)

265   Northern cassinia flower-
head, close up, Cape Reinga
(January)

264–265   Northern cassinia, *Cassinia amoena,* is a
low shrub, about 80 cm high (fig. 264), with thick,
fleshy leaves that are covered below with a dense
white tomentum. It is found only in scrub on cliff
faces near North Cape, and flowers (fig. 265) occur
in corymbs, about 3 cm across, during January and
February.   ASTERACEAE

266   Totorowhiti/grass tree/turpentine shrub,
Dracophyllum strictum, is found in damp lowland
and subalpine scrub from Thames south to Nelson
and Marlborough from sea-level to 800 m. The
leaves, 3.5–10 cm long by 7–12 mm wide, have finely
serrulate-crenulate margins. Flowers in terminal
panicles, 5–10 cm long, occur during April and May.
EPACRIDACEAE

266   Totorowhiti flower spray, Nelson (April)

267   Niniao, *Helichrysum glomeratum,* is a much-
branched shrub about 3 m high with interlacing
branches and rounded leaves, 10–25 mm long, on
flattened petioles about 5 mm long. It is found in
lowland scrub, along forest margins and sometimes
in rocky places throughout New Zealand. The small,
ball-like clusters of flowers, about 5 mm across,
occur from November to January.   ASTERACEAE

267   Niniao flower-heads, Lake Pounui (December)

268 Variable coprosma, *Coprosma polymorpha,* forms a spreading shrub with divaricating branches, pubescent branchlets and variable leaves, 10–25 mm long by 2–6 mm wide, varying from ovate to lanceolate to linear in shape. Drupes ripen during April and May, and the shrub is found in lowland scrub and forest throughout New Zealand, though often local in occurrence. ASTERACEAE

270 Shrubby kohuhu, *Pittosporum rigidum,* is a densely branched, rigid shrub, to 3 m high, with small, thick, leathery leaves, 8–10 mm long by 5–8 mm wide, found occasionally in scrub but more commonly from montane to subalpine regions from the Kaimanawa to the Southwest Nelson mountains. Small axillary flowers occur from September to February. PITTOSPORACEAE

268 Branchlet of variable coprosma with drupes and leaves, Rahu Saddle (April)

269 Ure Valley tree daisy, *Olearia coriacea,* is a 3 m high shrub with very thick, rounded, leathery leaves having a dense brownish white tomentum below, and is found in scrub along the Seaward Kaikoura Range. Flowers occur singly during March and April. ASTERACEAE

270 Branchlets of shrubby kohuhu with flowers and leaves, Kaimanawa Mountains (October)

271 Hokotaka, *Corokia macrocarpa,* forms a shrub, up to 6 m high, with narrow, leathery leaves, 4–15 cm long. Flowers, 10 mm across, occur in axillary racemes from October to December; the drupe, ripe during April and May, is dark red. It is found along forest margins and in scrub in the Chatham Islands. CORNACEAE

269 Flowering stem of Ure Valley tree daisy, Upper Ure River (March)

271 Hokotaka flowers, Otari (October)

272 Poroporo, *Solanum laciniatum,* is similar to *S. aviculare* (p. 40) except that *S. laciniatum* has distinctly purplish blue leaves and stems. It is found mainly in the North Island along forest margins and in scrub. SOLANACEAE

272 Poroporo flowers, Waikanae (November)

273 North Cape hebe, *Hebe macrocarpa* var. *brevifolia,* is found only around the North Cape as a stiffly branched shrub, about 2 m high, with spreading leaves, 6–12 cm long by 10–30 mm wide, and bearing these lovely flowers almost all the year round. SCROPHULARIACEAE

273 North Cape hebe flower-spike, North Cape (October)

274 Houpara leaves, Coromandel Peninsula (December)

274–275 Houpara, *Pseudopanax lessonii,* is one of three species of *Pseudopanax,* all of which have thick, leathery, unifoliolate or 3–5 foliolate leaves on long petioles. The leaves of *P. lessonii,* with their petioles 5–15 cm long, are shown in fig. 274, and the flower, which occurs from December to February, is shown in fig. 275. It is a shrub or small tree, up to 6 m high, and grows in lowland scrub and forest from the Three Kings Islands to Poverty Bay. ARALIACEAE

275 Houpara flower, close up, Coromandel Peninsula (January)

276–279 Toothed lancewood, *Pseudopanax ferox,* forms a small tree up to 5 m high, with a slender, 'roped' trunk (fig. 279) and long, toothed, upward-spreading, adult leaves, 5–15 cm long by 10–20 mm wide (fig. 277); leaves of juveniles are up to 50 cm long, and drooping (fig. 278). Flowers and fruits occur from January till April (fig. 276), and the tree is found in lowland scrub and forest from Mangonui southwards in the North and South Islands but is rather rare in occurrence. ARALIACEAE

276 Toothed lancewood berries, Otari (April)

277 Toothed lancewood, showing juvenile pendant leaves and adult ascending leaves, Otari (November)

278 Toothed lancewood leaves, Otari (November)

279 The 'roped' trunk of the toothed lancewood, Otari (September)

280 *Pseudopanax discolor* leaf from Kauaeranga Valley (November)

281 *P. discolor* leaf from near Thames (November)

280–281 ***Pseudopanax discolor*** is a shrub or small tree, up to 5 m high, that grows in lowland scrub and forest from Mangonui to about Thames and the Manukau Harbour. Trifoliolate and five-foliolate leaves on their petioles, 2–8 cm long, are shown in figs 280 and 281 respectively. ARALIACEAE

282 ***Pseudopanax gilliesii*** is a shrub or tree up to 5 m high, with slender branches and unifoliolate or trifoliolate, and sometimes irregular-lobed, leaves, 4–8 cm long on long, slender petioles up to twice as long as the leaves (fig. 282). It is found growing in scrub only near Whangaroa and on Little Barrier Island. ARALIACEAE

282 *Pseudopanax gilliesii* leaf spray showing the very long leaf petioles, Whangaroa (October)

PLANTS OF BOGS AND SWAMPS

Plants that grow in these situations are tolerant of continuous wet conditions around their roots. They are found in swampy places in both lowland and alpine regions and often occur round the seepages of springs in bush-clad gullies or on mountainsides. Bog conditions of only a few square metres in area often provide ideal places in which these plants can thrive.

283 Blue swamp orchids in flower, Te Anau (January)

283 Blue swamp orchid, *Thelymitra venosa*, is a lovely ground orchid, 20–50 cm high, found in swamps throughout New Zealand. Flowers appear during December and January. ORCHIDACEAE

284 Raupo, *Typha muelleri*, is a bullrush, up to 2.7 m high, found in swamps and marshes throughout New Zealand. The leaves are up to 2.5 cm wide, and flowers occur from December through March. TYPHACEAE

284 Raupo in flower on edge of Lake Pounui (January)

285 Flowers and flower-buds of Oliver's dracophyllum, Lake Manapouri (November)

285 Oliver's dracophyllum, *Dracophyllum oliveri*, is a species of *Dracophyllum* that grows to 1 m high and is found in swamps near Te Anau. Flower racemes occur along the branches from November to January. EPACRIDACEAE

286 Toetoe, *Cortaderia toetoe*, is New Zealand's largest endemic grass, found throughout the country in lowland swampy places and along riverbanks, and on sandhills. The plume-like flowers, up to 3 m high, occur from November through March.

GRAMINEAE

287 Flax plant in flower,
Waiorongomai (December)

288 A bronze-leaved form of flax,
Taupo (August)

286 Toetoe plant in flower, Lake
Pounui (December)

287–288 Flax/harakeke, *Phormium tenax*, is one of New Zealand's most handsome plants. It is common in lowland swamps throughout the country and on the Kermadec and Chatham Islands. Leaves are up to 3 m long by 5–12.5 cm wide and the flowers stalks reach 5 m high; the flowers, each about 2.5 cm long, appear along the stalks from November to January. Several cultivars with richly coloured leaves have been developed (fig. 288).

PHORMIACEAE

289 Sweet-scented grass, Hinakura (February)

289 Sweet-scented grass, *Hierochloe redolens*, is an erect, strongly sweet-scented grass, up to 1 m high, common around boggy places throughout New Zealand. The flower panicles, 35 cm long, appear from December through February. GRAMINEAE

290-291 Soft herb, *Myriophyllum pedunculatum*, is a soft aquatic herb with creeping, rooting stems producing simple, upright branches up to 10 cm high (fig. 290). Fig. 291 shows the flowering patch made by this tiny plant growing amongst a gunnera. Flowers, 4 mm across, occur from October to February, with female flowers at the branch apex and male flowers just below (fig. 290). Found in peaty, wet places from Mangonui southwards to Stewart and Chatham Islands.   HALORAGACEAE

291   Soft herb growing among solitary gunnera and flowering, Boulder Lake (February)

290   Soft herb plants in flower, Boulder Lake (February)

292 Kuwawa, *Scirpus lacustris*, is a leafless sedge, up to 2 m high, found along swamp and lake margins throughout New Zealand. Flower-heads, 5–10 cm long, occur from November to February, and the fruits ripen by May.   CYPERACEAE

292   Kuwawa in fruit, Lake Pounui (May)

293 Wi, *Juncus pallidus*, is a tall, dense, tufted herb, 1.75 m high, found in lowland swampy places throughout New Zealand. Large cymes of flowers, each 3 mm across, occur from December to February.   JUNCACEAE

293   Flowers of wi, near Rotorua (February)

294 Cutty-grass, *Carex geminata*, is a robust sedge, 50 cm–1.3 m high, with flat, scabrid-margined leaves, 5–12 mm wide, which can cut the skin if grasped. It is found throughout New Zealand on the edges of swamps, along streamsides and in damp places in the forests. The flower-spikes, up to 10 cm long, appear from November to February.

CYPERACEAE

294 Cutty-grass with flower-spikes, Mt Tauhara (January)

295 Dense nertera, *Nertera balfouriana*, forms dense patches of a creeping plant up to 25 cm across in damp or boggy ground, often among sphagnum moss, from the Kaimanawa Range southwards, at elevations between 600 m and 1,000 m. The leaves and flowers are small but the brilliant drupes, 7–9 mm long, are ripe during February and March.

RUBIACEAE

295 Dense nertera plant with drupes, Volcanic Plateau (February)

296 Alpine cushion, *Donatia novae-zealandiae*, is a small, firm, tufted plant, which forms hard cushions in alpine bogs from the Tararua Range southwards to Stewart Island, where it descends to sea-level. It flowers profusely during January and February, with small flowers 8–10 cm across.

DONATIACEAE

296 Alpine cushion plant in flower, Arthur's Pass (January)

297 Prostrate grass tree, *Dracophyllum muscoides*, is a creeping grass tree with stout, dark, brownish black stems, up to 30 cm long, found around sub-alpine bogs, damp grasslands and herbfields of the South Island from Lake Ohau southwards. Branchlets are clothed with thick, leathery, imbricating leaves about 6 mm long by 2.5 mm wide; the flowers appear during January and February.

EPACRIDACEAE

297 Prostrate grass tree with flowers, Key Summit (January)

298 Peat swamp orchids in flower,
Moana Tua Tua Bog (September)

299 Alpine sundew, *Droscera arcturi*, is an insectivorous plant found in bogs and swamps from the
Volcanic Plateau to Stewart Island. The narrow,
strap-like leaves with their petioles are 5–12 cm long
and bear sticky hairs that, as in all sundews, entangle
insects, which the plant consumes to obtain its
nitrogen. Flowers 16 mm across on stalks 15 cm high
occur from November to January. DROSCERACEAE

299 Alpine sundew plants in flower, Key Summit
(January)

298 Peat swamp orchid, *Corybas unguiculatus*, is
one of the rarest New Zealand plants, found only
in wet, peaty swamps of the northern North Island.
Flowers appear in mid-July, at first as horizontal
buds, which continue growing in size and length of
stalk until September or early October, when they
open. Unless pollinated, flowers collapse and wither
within ten days. Flowers are 2 cm long on stalks
10–20 mm high.   ORCHIDACEAE

300 Spathulate sundew, *Droscera spathulata*, is a
small insectivorous plant, usually occurring in groups
in lowland and subalpine bogs and swamps throughout New Zealand. The glandular leaf petiole, 10 mm
long, widens to a spoon-shaped lamina, 5 mm long,
bearing stalked, glandular hairs. Flower-stalks, each
carrying racemes of up to 15 white or pink-coloured
flowers, occur from November till January.
                                  DROSCERACEAE

300 Spathulate sundews, Key Summit (January)

301 Scented sundew, *Droscera binata*, has upright
leaves, 15 cm long by 2 mm wide, bearing glandular
hairs and arising in twos and threes from petioles
up to 35 cm long. Flower-stalks up to 50 cm high
bear white flowers, 2 cm across, in cymes from
November to February. Scented sundew is found in
lowland and montane bogs and swamps throughout
New Zealand.   DROSCERACEAE

301 Scented sundews, Lake Te Anau (January)

302   Hollow-stemmed sedge with
flower-spikelets, Lake
Pounui (January)

303   Wahu showing spathulate tip
of leaf, Renata Peak, Tararua
Mountains (November)

302   Hollow-stemmed sedge, *Eleocharis sphacelata,*
is a sedge with hollow, tubular stems up to 1 m high,
found throughout New Zealand in wet swamps, and
along the margins of lakes from sea-level to 800 m.
The terminal flower-spikelets, white when fresh, are
up to 5 cm long.   CYPERACEAE

303–304   Wahu, *Droscera stenopetala,* is a variable
sundew, normally with narrow, spathulate leaves,
2 cm long by 5–6 mm wide on stout petioles 5 cm
long by 8 mm wide (fig. 304). The leaves (fig. 303)
have long, sticky, glandular hairs. Solitary white
flowers, 2 cm across on peduncles 20 cm long, occur
from November to January. This wahu is found in
montane and subalpine bogs and swamps from the
Ruahine Range south to Stewart Island, coming
down to sea-level in the far south.

DROSCERACEAE

304   Wahu plant, Renata Peak, Tararua Mountains
(November)

305   Moss daisy, *Abrotanella caespitosa,* is a rather
moss-like plant, with thick, leathery leaves,
10–15 mm long by 1–1.5 mm wide, which form
rounded matted patches, up to 10 cm across, along
the edges of subalpine bogs and seepages in grasslands and herbfields from the Ruahine Range south
to northern Fiordland. Minute flowers occur during
January and February.   ASTERACEAE

305   Moss daisy with flower-buds, Renata Peak,
Tararua Mountains (November)

306–308 Niggerhead, *Carex secta*, forms a large sedge up to 1.5 m high whose matted roots and decaying leaves form huge, broad pillars, 1–1.5 m high (fig. 306), which lift the sedge up to form the conspicuous objects in swamps called 'niggerheads' (fig. 308). The leaves are 1–2 m long by 3–4 mm wide, flat above, keeled below, with sharp cutting margins. Flower panicles (fig. 307), up to 1 m in length, occur from December to January. Niggerheads are found throughout New Zealand in swamps, bogs, along stream banks and in other damp places.

CYPERACEAE

306 Niggerhead in flower, Lake Lyndon (January)

307 Flower panicles of niggerhead, Lake Lyndon (January)

308 Niggerheads in bog near Garston (January)

309 Horizontal orchid, *Lyperanthus antarcticus*, is an unusual orchid, with its flower held horizontally, almost at right angles to the stem. The leaves are 2.5–7 cm long and the flowers, which occur from November to March, 8–14 mm long. It is found in subalpine bogs and swamps from the Tararua Range to Stewart Island.

ORCHIDACEAE

309 Horizontal orchids flowering, Key Summit (January)

310 Bog daisy, *Celmisia graminifolia*, is a small daisy found throughout the country on the edges of montane and subalpine bogs and swamps and in damp places in alpine grasslands, herbfields and fellfields. The pointed leaves, 5–20 cm long by 4–5 mm wide, are clothed below with dense felt; the flowers, 10–15 mm across, occur during December and January.   ASTERACEAE

310   Bog daisies in flower, Key Summit (January)

311 Small alpine swamp sedge, *Carex flaviformis*, is a short-leaved sedge found in swamps in mountain regions of the South Island between 500 and 1,200 m altitude. Tufted flower-heads arise from December to February.   CYPERACEAE

311   Small alpine swamp sedge
with seed-heads, Key Summit
(January)

312 Native oxalis, *Oxalis lactea*, is one of three species of *Oxalis* found near bogs, swamps and damp places alongside streams in lowland and subalpine regions throughout New Zealand. It spreads by creeping rhizomes, and the patches of three-lobed leaves can reach 20 cm across, with the flowers, 12–20 mm across, occurring from October to March.
                                OXALIDACEAE

312   Native oxalis in flower, Opotiki (November)

313 Turf-forming astelia, *Astelia linearis*, and *A. subulata* are two species of turf-forming *Astelia* found in alpine wet and boggy places in the South Island mountains from the Paparoa Range to the Longwoods. The leaves are 5–10 cm long by 2–6 mm wide, and the plant forms extensive patches.
                                ASPHODELACEAE

313   Turf-forming astelia with seeds, Upper Stillwater
(December)

PLANTS OF STREAMSIDES, DAMP OR SHADY PLACES

Plants that grow in these situations prefer cool, moist, but not wet, root runs. They are found from lowland to alpine regions, growing on banks, either shaded by trees and shrubs or open and facing away from the hot noonday sun. They also grow on the drier, but still damp, ground above the edges of bogs and swamps and in the cool alluvium of river and streamsides. They are found, as well, in rocky places kept damp by seepages and rivulets.

314 Creeping gunnera, *Gunnera prorepens*, is a creeping plant, forming patches 40 cm or more across in lowland and subalpine damp or boggy places throughout New Zealand, and is often found in sphagnum moss growing along the edges of bogs. The leaves are about 3 cm long and the flowers appear on 6 cm high stalks from September to January; the drupes, 3–4 mm long, ripen from February on. HALORAGACEAE

314 Creeping gunnera plant with ripe drupes, Outerere Stream, Volcanic Plateau (February)

315 Red-fruited gunnera plant, Travers Valley (April)

315 Red-fruited gunnera, *Gunnera dentata*, is a creeping plant, forming mats up to 60 cm across along subalpine streamsides and in damp places throughout New Zealand; sometimes several plants may coalesce to form one large mat 2–3 m across. The dentate leaves are 8–15 mm long on petioles up to 5 cm long; the flowers appear in November and December, and the pendulous drupes (fig. 315), occur as open clusters on elongated stalks during March and April. HALORAGACEAE

316–317 Solitary gunnera, *Gunnera monoica,* is a creeping mat plant found up to 1,000 m altitude on damp, shaded clay banks or damp, stony river-beds, from about Thames southwards. The dentate-serrate leaves, 10–15 mm long, are on hairy petioles up to 4 cm long (fig. 317). The flowers (fig. 316) appear during October and November on erect stalks up to 7 cm high, with male flowers occupying the upper three-quarters of the stalk and female flowers the lower quarter. Ripe drupes appear from December to February.   HALORAGACEAE

317   Plant of solitary gunnera with immature drupes, Boulder Lake (February)

316   Solitary gunnera with flower-stalks, Travers Valley (November)

318   Swamp musk, *Mazus radicans,* is a creeping perennial herb, found between 200 m and 1,200 m, in damp places and along the edges of bogs or swamps from the Ruahine Range to Otago. The narrow, obovate leaves, with their petioles, are hairy and 2–5 cm long. The flowers occur from November to March. A similar plant, *M. pumilio,* has longer, obovate-spathulate leaves that are less hairy.   SCROPHULARIACEAE

318   Swamp musk in flower, Travers Valley (December)

319   Purple bladderwort, *Utricularia monanthos,* is a plant that drops its leaves before it flowers and has small bladders on its roots. These bladders, 1.5–2.5 mm wide, have opening lids and trap microscopic water animals, which are digested to provide nitrogen for the plant. The flowers are about 10 mm across on stems 10 cm high and occur during January. Found in bogs and swamps from the Kaimanawa Range southwards.

LENTIBULARIACEAE

319   Purple bladderwort flowers, Key Summit (January)

320 *Carex dissita* forms a handsome, tall, tufted sedge, up to 1 m high, found in abundance in damp areas of lowland and subalpine forests throughout New Zealand. The broad, flat, dark green leaves are deeply grooved, and the flower-heads appear from September to January. CYPERACEAE

320 *Carex dissita* leaves with flower- and seed-heads, near Taupo (January)

321 Parataniwha, *Elatostema rugosum*, is a fleshy, decumbent and spreading plant found in shaded, damp places, especially by streamsides, in lowland forests from about North Cape to the Tararua Ranges. The attractive, alternate, coarsely serrate and rather rough leaves are 8–25 cm long by 2.5–6 cm wide. URTICACEAE

321 A parataniwha plant displays its colourful leaves, Hukutaia Domain (April)

322 Puatea, *Gnaphalium keriense*, is a small, hanging or prostrate everlasting daisy, up to 24 cm high, with spreading branches and sessile leaves, 4–7 cm long; those towards the branch tips are shorter. Flowers arise as flat corymbs, each flower about 1 cm across, from September to February. Found from Northland to North Canterbury in lowland and subalpine regions along streamsides and shaded banks, often on shady roadside banks. ASTERACEAE

322 Puatea in flower, near Taihape (November)

323 Maori calceolaria, *Jovellana sinclairii*, found from the East Cape southwards in damp places, is a small herb, up to 30 cm high, with upright, downy stems and thin, opposite leaves about 8 cm long. Flowers occur as branched panicles from October till February. SCROPHULARIACEAE

323 Maori calceolaria flowers, Pahaoa River Valley (December)

324   Red spider orchid, *Corybas macranthus,* is a small, low-growing orchid with thin leaves, 5 cm long, found throughout New Zealand from sea-level to 700 m on clay banks and logs and among moss on logs in damp, shaded places. The spider-like flowers occur during October and November. When the seeds have set, the stalks elongate to about 10 cm so that the seeds may be borne away by wind.

ORCHIDACEAE

324   Red spider orchids on a bank near Hinakura (November)

326–328   Lantern berry/puwatawata, *Luzuriaga parviflora,* is a delicate creeping lily found throughout New Zealand in lowland forests, growing on damp, moss-covered banks and moss-covered tree trunks and logs. Leaves, 10–27 mm long by 3–6 mm wide, are alternate, and flowers, 15–35 mm across, occur from November to February, with berries appearing from January to March.

LILIACEAE

325   Close-up of flowers of cudweed, Haast (November)

325   Cudweed, *Gnaphalium hookeri,* is a trailing everlasting daisy with alternate leaves, 5–10 cm long, and flowers, 15 mm across, occurring in corymbs 7.5–10 cm across, from November to January. Found on damp, shady banks in lowland and subalpine regions from Taupo southwards.

ASTERACEAE

327   Lantern berry leaves and flower, Mt Ngamoko (January)

326   The flower of lantern berry, Mt Ngamoko (January)

328   Lantern berry, spray with berry, Mt Ngamoko (January)

329 Yellow-leaved sedge, *Carex coriacea*, forms a small, yellow-green-leaved sedge, 50–100 cm high, found in damp grassland, swampy seepages or swampy river flats from sea-level to 1,200 m. Leaves are longer than the culms; flowers occur during January, and fig. 329 shows a plant with male flowers on the tip of a spike and female flowers below. The shoots of this sedge die back completely during winter. CYPERACEAE

329 Yellow-leaved sedge with flower- and seed-heads, Lake Te Anau (January)

332 Scented broom, *Carmichaelia odorata*, is a branching shrub, up to 3 m high, with drooping, compressed, grooved and striated branchlets up to 20 cm long. Scented flowers occur as 15–20-flowered racemes during October and November and, often again, during February and March. Found along shady banks, streamsides and forest margins from Lake Waikaremoana southwards. FABACEAE

332 Scented broom in flower, my garden at Waikanae (October)

330–331 Tree broom, *Chordospartium stevensonii*, is a small, leafless canopy tree, up to 8 m high, with pendulous branchlets that bear 9 cm long racemes of mauve flowers in profusion from November to January. The tree is found on isolated alluvial river and stream terraces in the vicinity of the Clarence, Awatere and Wairau rivers. FABACEAE

330 Tree broom flowers, Jordan River (December)

331 Flowers of tree broom, close up, Jordan River (December)

334   Plant of yellow snow marguerite in full flower, Homer Saddle (January)

333   Flowers of yellow snow marguerite, Arthur's Pass (January)

333–334   Yellow snow marguerite, *Dolichoglottis lyallii*, forms a plant 50 cm high with slender stems terminating in corymbs of brilliant yellow flowers from December to February (fig. 334), each flower 4–5 cm across (fig. 333). The daffodil-like leaves are up to 25 cm long by 10 mm wide, and the plant is found in damp rocky places, and damp situations in grasslands, herbfields and fellfields along the Southern Alps, and in Fiordland and Stewart Island.   ASTERACEAE

335   Flowers of snow marguerite, Arthur's Pass (January)

335–336   Snow marguerite, *Dolichoglottis scorzoneroides*, forms a plant reaching 60 cm high with stout stems and broad, pointed leaves up to 20 cm long by 2–3 cm wide (fig. 336). The snow marguerite bears broad corymbs of white flowers from December to February, each flower 4–6 cm across (fig. 335). It is found in similar situations to *D. lyallii*, and these two species readily hybridise with one another, producing plants of varying form and flower colour.   ASTERACEAE

336   Plant of snow marguerite in flower, Homer Saddle (January)

337 Flowers and leaves of large-
flowered pink broom,
Milford Sound (December)

337–338 Large-flowered pink broom, *Carmichaelia (Thomsoniella) grandiflora* is a branching, spreading shrub up to 2 m high, bare in winter but leafy in spring and summer, and bearing 5–10-flowered racemes of fragrant flowers (fig. 337), each 6–8 mm long, during December and January. The seed-pods shown in fig. 338 illustrate the typical apical, beak-like structure of native broom seed-pods. Found along streamsides and in shaded places in the South Island west of the Southern Alps.

FABACEAE

338 Seeds of large-flowered pink broom, Milford Sound (January)

339 Leafy broom, *Carmichaelia angustata,* forms a shrub or small tree reaching 2 m high, with grooved, flattened branches and occasional 3–7-foliolate leaves. Flowers in 10–40-flowered racemes occur from February to April. Found west of the Southern Alps along streamsides and montane forest margins.

FABACEAE

339 Leaf spray of leafy broom, Otari (April)

340   Stems and flowers of South
        Island broom, Milford Track
        (April)

341   Flowers of South Island
        broom, Milford Track
        (April)

340–341 South Island broom, *Carmichaelia arborea*, forms a tree up to 5 m high with ascending branches (fig. 340) and closely set, compressed, striated branchlets. Flowers in 3–5-flowered racemes (fig. 341) occur from February to April. Found only in the South Island, west of the Southern Alps, along lowland streamsides and forest margins, in scrub and on alluvial swampy and boggy ground. FABACEAE

342 Common nertera, *Nertera depressa*, forms patches up to 30 cm across on damp, shady banks, in forest and scrub, on the edges of bogs and in damp grasslands and herbfields in lowland and subalpine regions throughout New Zealand. Minute flowers occur from November to February, and red drupes ripen from January to May.   RUBIACEAE

343–344 Broad-leaved sedge, *Vincentia sinclairii*, is a tall, leafy sedge, with broad leaves up to 1.3 m long, found on shaded cliffs or banks and along stream and lakesides from North Cape to the Kaimanawa Mountains. The flowers (fig. 343) occur from October to January, followed by the large, conspicuous seed-heads (fig. 344).   CYPERACEAE

342   Common nertera with drupes, Hollyford Valley
        (April)

343   Broad-leaved sedge with
        flower-head, Waipunga
        Gorge (October)

345–346 New Zealand violet, *Viola cunninghamii,* is a small, tufted herb (fig. 345), up to 15 cm high, found in damp places in lowland and subalpine regions throughout the country. Flowers (fig. 346), 10–20 mm across, on stalks up to 10 cm long, occur from October to January. The leaves, 3 cm long, are on petioles 1–10 cm long.   VIOLACEAE

346 Close view of New Zealand violet, *V. cunninghamii,* flowers, Dun Mountain (November)

345 New Zealand violet, *Viola cunninghamii,* plant in flower, Thomas River (December)

347 New Zealand violet, *Viola lyallii,* forms small plants about 15 cm high with stems that creep only a short distance before turning upwards at their tips. The cordate leaves, 3 cm long, are on petioles up to 10 cm long and the flowers, 10–20 mm across, are without any scent and occur from October to January. Found in damp, shady places from sea-level to 1,200 m throughout New Zealand. VIOLACEAE

344 Broad-leaved sedge with seed-heads, Waitahanui Stream, Taupo (December)

347 New Zealand violet, *Viola lyallii,* flower sprays with leaves, Wharite Peak (December)

348   Thistle-leaved senecio, *Senecio solandri* var. *rufiglandulosus*, is a leafy shrub up to 1 m high, with deeply lobed, dentate leaves 20 cm long. Flowers, 2 cm across, occur in corymbs from November to February. Found along damp, shaded banks and streamsides in lowland and subalpine regions from Mt Taranaki southwards.   ASTERACEAE

348   Thistle-leaved senecio plant in flower, Mt Egmont (January)

349   Native pin cushion, *Cotula squalida*, forms a creeping plant, up to 40 cm across, with serrated, fern-like leaves, 2.5–5 cm long. Flowers, like small pin-cushions, 6 mm across, occur in profusion from December to February. Found in damp grassland and rocky situations in lowland and alpine regions all over New Zealand.   ASTERACEAE

349   Native pin cushion plant in flower, Lewis Pass (December)

350   Mountain cotula plant in flower, Cupola Basin (December)

350   Mountain cotula, *Cotula pyrethrifolia*, is a very aromatic herb, with creeping and rooting stems bearing pinnatifid leaves. Flowers are 8–20 mm across and occur on long stems from October to January. Found throughout New Zealand, between 800 and 2,000 m, along streamsides, and in damp gravel, grassland and herbfields.   ASTERACEAE

351   Feathery-leaved cotula plant in flower, Waitahanui Stream (February)

351   Feathery-leaved cotula, *Cotula minor* is a creeping, rooting plant with silky-hairy stems up to 40 cm long bearing many thin, pinnate and pin-natisect leaves, 10–50 mm long by 5–10 mm wide. Small flowers, 2–5 mm across, occur from December to February. Found in lowland to montane regions on wet ground along the margins of swamps, stream-sides, shaded grassy places and the edges of sandy tidal estuaries.   ASTERACEAE

352 Leafy forstera, *Forstera bidwillii* var. *densifolia,* is a herb with densely leafy stems, the leaves 12 mm long by 6 mm wide on stems 20 cm long. Flowers, 8–15 mm across, occur on long stems as 1–3-flowered clusters from December to March. *F. bidwillii* is found in subalpine and alpine regions, growing on damp rock faces or in damp grasslands and herbfields. The variety *densifolia* grows only on Mt Taranaki. STYLIDIACEAE

352 Leafy forstera plant in flower, Mt Taranaki (January)

353 Slender forstera, *Forstera tenella,* is a smooth-leaved, herbaceous plant found in alpine regions from the Ruahine Range southwards. It grows in damp grasslands, bogs and herbfields, and flowers, up to 10 mm across, occur from December to February. STYLIDIACEAE

353 Flowers of the slender forstera, Lewis Pass (January)

354 Flowers of panakenake, Mt Taranaki (December)

355 Panakenake plant with berries, Arthur's Pass (April)

354–355 Panakenake/creeping pratia, *Pratia angulata,* is a slender, creeping herb, forming mats up to 1 m across and found in damp, sheltered places up to 1,500 m altitude. The flowers (fig. 354), 8–16 mm across, occur in profusion from October to March. The berries, 8–12 mm long (fig. 355), ripen from February to April. LOBELIACEAE

356 Chatham Island pratia, *Pratia arenaria*, forms patches 1 m across in sandy, damp places throughout the Chatham Islands, in south-east Otago, and on the Antipodes Islands. The leaves are 10–15 mm long, the flowers 10–12 mm across and the berries 7–10 mm long. Flowers occur from December to March and ripe berries from February to May.

LOBELIACEAE

356 Chatham Island pratia plant in flower, Otari (February)

358 Flowering plant of creeping ourisia var. *gracilis*, Homer Cirque (December)

357 Close-up of creeping ourisia flowers, Travers Valley (December)

357–358 Creeping ourisia, *Ourisia caespitosa* var. *caespitosa* (fig. 357), is a mat-forming plant, with leaves 10 mm long by 5 mm wide found in damp, rocky places from the Ruahine Range southwards. Fig. 358 shows the variety *gracilis*, which has shorter leaves, 6 mm long by 3 mm wide, and which is found in damp places only in the mountains of Canterbury and Otago. Flowers of *gracilis* occur mainly in pairs while those of *caespitosa* occur mostly singly.

SCROPHULARIACEAE

359 Sand ranunculus, *Ranunculus acaulis*, is a low-growing, creeping plant with tufts of fleshy, 3-folio-late to deeply three-lobed leaves, up to 2 cm long, on petioles 2.5–5 cm long. Flowers, 6–9 mm across, occur from September through November, and the plant is found in damp, sandy places along the coasts or along damp, sandy shores of inland lakes.

RANUNCULACEAE

359 The sand ranunculus in flower, Opotiki coast (November)

360-361   North Island mountain foxglove, *Ourisia macrophylla*, is a perennial herb up to 60 cm high, which forms colonies from a creeping rhizome. The broad, dentate leaves are up to 15 cm long, and the whorls of flowers (fig. 361), with each flower about 2 cm across, occur from October to February. Found only in the mountains in damp, shaded places in subalpine scrub and herbfields and along shaded banks and streamsides. Several varieties are known from different localities (fig. 360).

SCROPHULARIACEAE

360   North Island mountain foxglove plants, var. *meadii*, in flower, Mangakino (October)

361   Whorls of flowers of North Island mountain foxglove, Mt Taranaki (December)

363   Hairy ourisia, *Ourisia sessilifolia* var. *splendida*, is a similar plant to *O. sessilifolia* var. *simpsonii* but with the leaves and stems much more heavily clothed with long, soft hairs. Flowers, 15–20 mm across, on hairy stalks up to 15 cm long, occur from December to February. Found in damp situations in the mountains from north-west Nelson to Fiordland.

SCROPHULARIACEAE

362   Hairy ourisia, *Ourisia sessilifolia* var. *simpsonii*, forms a small plant with hairy leaves formed as rosettes along a creeping stem. Flowers in pairs, each 10–15 mm across, occur from December to February on hairy stems 5–15 cm high. Found in the South Island mountains in damp grassland and wet fellfields up to 2,000 m altitude.

SCROPHULARIACEAE

362   Hairy ourisia plants in flower, Cupola Basin (December)

363   Hairy ourisia, var. *splendida*, in flower, Wapiti Lake, Fiordland (December)

364 South Island mountain
foxglove plant in flower,
Temple Basin (January)

364 South Island mountain foxglove, *Ourisia macrocarpa* var. *calycina*, is a similar plant to *O. macrophylla* and is found in damp scrub and herbfields and along streamsides in the South Island mountains from Nelson to Fiordland. Flower stalks up to 70 cm high bear flowers from October to January.

SCROPHULARIACEAE

365 *Potentilla anserinoides* is a woody, prostrate plant with masses of pinnate leaves, up to 15 cm long, each having a single terminal pinna. The plant is found in lowland to montane damp grasslands and on the edges of bogs from Thames southwards. Flowers occur from September to February.

ROSACEAE

365 *Potentilla anserinoides* plant with flower,
Waipahihi Botanical Reserve (February)

366 Small New Zealand gentian, *Gentiana grisebachii*, is an annual with weak, ascending stems, 7–20 cm across, and thin, spathulate leaves, 15–20 mm long by 8–10 mm wide. Solitary or paired flowers about 12 mm long occur in January and February. Found in damp places in herbfields, subalpine scrub and grasslands throughout New Zealand.

GENTIANACEAE

366 Flower of small New Zealand
gentian, Mt Ruapehu (February)

367 Maori hypericum plant in flower,
The Wilderness (January)

367 Maori hypericum, *Hypericum japonicum*, is a procumbent, much-branched plant, forming patches up to 20 cm across and bearing small yellow flowers, singly, during November and December. Found in open damp places, along edges of tarns and slow-flowing seepages throughout New Zealand.

HYPERICACEAE

368 Wet rock hebe, *Parahebe linifolia,* is a branching shrub, forming a mat up to 40 cm across in wet rocky places and alongside streams between 800 and 1,400 m in the South Island mountains. The thick, sessile, narrow leaves are 8–20 mm long by 1.5–4 mm wide, and the flowers appear in 2–4-flowered racemes during December and January.

SCROPHULARIACEAE

368 Wet rock hebe plant in flower at Arthur's Pass (December)

369 Small-leaved wet rock hebe, *Parahebe lyallii,* forms a prostrate, spreading shrub, carpeting wet rocks and streamsides from 800 to 1,400 m in the mountains from the Ruahine Range to Fiordland. The thick, fleshy leaves are 5–10 mm long by 4–8 mm wide, and the racemes of flowers appear on long stalks, 8 cm long, from November to March.

SCROPHULARIACEAE

369 Spray of small-leaved wet rock hebe with flowers, Boulder Lake (January)

370 Flower spray and leaves of streamside hebe, Tararua Ranges (December)

370 Streamside hebe, *Parahebe catarractae,* is a low, open-branched shrub with serrated leaves, 10–40 mm long by 5–20 mm wide, found in sub-alpine regions throughout New Zealand along damp streamsides, cliff faces and wet, rocky places. Flowers, 8–12 mm across, are borne profusely as few- or many-flowered racemes on long stalks from November to April.

SCROPHULARIACEAE

371 Tangle herb with flowers and seeds, Lake Rotoiti, Nelson (December)

371 Tangle herb, *Rumex flexulosus,* is a peculiar tangled plant found throughout New Zealand in damp, rocky places and damp grasslands. The leaves may reach 30 cm in length. Tiny flowers appear in clusters from November to March, and fruits about 1 mm long ripen from January to April. The stems are grooved, and both stems and leaves are always brown in colour.

POLYGONACEAE

372 *Ranunculus godleyanus* grows to 60 cm high, with rounded crenate leaves up to 18 cm long. Yellow flowers, 2–5 cm across, occur on long stalks from January to March, and the plant is found in wet, subalpine, rocky places in the South Island from about Mt Travers to the Ben Ohau Range.

RANUNCULACEAE

372 *Ranunculus godleyanus* plant, Waterfall Valley, Cass River (December)

373 Star herb, *Libertia pulchella*, is a small, dainty plant about 12 cm high, with grass-like leaves having the lower surface duller than the upper. The flowers, 15 mm across, occur from November to February, and the rounded, smooth seeds are yellow when ripe.

IRIDACEAE

373 Star herb plant in flower, Wharite Peak (December)

374 Creeping lily plants in flower, The Wilderness (December)

375 Close-up of creeping lily flower, The Wilderness (December)

374–375 Creeping lily, *Herpolirion novae-zelandiae*, is a creeping plant with grass-like leaves, making patches up to 1 m across. The flowers, 2–3 cm across, may be white, pale blue or pale mauve and occur during January and February. This is one of the smallest lilies in the world; it is found in damp places and along the edges of bogs in lowland and subalpine regions, up to 1,250 m, throughout New Zealand.

LILIACEAE

376 Small feathery-leaved buttercup, *Ranunculus gracilipes*, with its finely pinnate or divided leaves on hairy petioles 2–12 cm long, is a delicate little herb, about 15 cm high, found in damp grasslands and herbfields alongside streams in the South Island from Mt Travers southwards. The flowers, 2 cm across, occur singly from November to December.

RANUNCULACEAE

376 Small feathery-leaved
buttercup plants in flower,
Cupola Basin (December)

377 False buttercup, *Schizeilema haastii*, looks like a buttercup but on closer examination will show tiny umbelliferous flowers. The plant is found alongside water seepages from steep rock faces and among rocks in fellfields between 1,000 and 1,400 m from the Ruahine Range southwards.   APIACEAE

377 False buttercup with flowers, Homer Cirque
(January)

378 Forest floor lily, *Arthropodium candidum*, grows as small colonies on the forest floor and other shaded places throughout New Zealand. The leaves are 10–30 cm long by 3–10 mm wide, and the flowers, about 10 mm across, occur from November to January.   ASPHODELACEAE

378 Forest floor lily in flower,
Wharite Peak (December)

379 Common willow herb, *Epilobium nerterioides*, is widespread in lowland and montane regions throughout New Zealand, growing in open, damp places and stream beds. It forms patches up to 20 cm across; the fleshy leaves are crowded, and the flowers, 2–3 mm across, occur from September to April.   ONAGRACEAE

379 Common willow herb in flower, Lake Tekapo
(October)

380–382　Mountain astelia, *Astelia nervosa,* is a tussock-like plant (fig. 380) with large, silvery leaves, 50 cm to 1.5–2 m long by 2–4 cm wide, covered by scales that become ruffed up to form a white fur. This lily can form large colonies covering extensive areas on mountainsides all over New Zealand in damp grasslands, herbfields and fellfields between 700 and 1,400 m. The flowers (fig. 382) are strongly sweet scented and appear from November to January, and the berries, about 8 mm across (fig. 381), ripen from March to May.　　　　ASPHODELACEAE

380　Mountain astelia plant growing on Mt Arthur (January)

381　Mountain astelia berries, Mt Holdsworth (May)

382　Mountain astelia flowers, Dun Mountain (November)

383　Kahaka/fragrant astelia, *Astelia fragrans,* is a tufted plant, growing either singly or in colonies. Leaves, 50 cm to 2–2.5 m long by 2.5–7.5 cm wide, are stiff and ascending in the lower half but flaccid in the upper half, usually with a strong red-coloured costa on either side of the midrib. Flowers occur in October and November, with berries ripening from December to May (fig. 383).　　ASPHODELACEAE

383　Kahaka showing berries and red costa on each side of leaf midrib, Mt Hector, Tararua Range (March)

384   Streamside tree daisy in full
       flower, Arthur's Pass (December)

385   Close-up of flowers of
       streamside tree daisy,
       Arthur's Pass (December)

384–385   Streamside tree daisy, *Olearia cheesemanii*, is a branching shrub, up to 4 m high, found along lowland and montane streamsides near forest margins, from Auckland south to about Arthur's Pass. The leaves, 5–9 cm long by 2–3 cm wide, are pale buff coloured below with a fine tomentum. Corymbs of flowers, up to 15 cm across, occur in profusion from October to January.

ASTERACEAE

386   Large white-flowered daisy plant in full flower,
       Takaka Hill (January)

387   Close-up of large white-flowered
       daisy, Takaka Hill (January)

386–387   Large white-flowered daisy, *Brachyglottis hectori*, is a shrub or small tree, up to 4 m high, with stout, brittle, spreading branchlets bearing oblanceolate leaves, 10–25 cm long by 4–12 cm wide, on 4 cm long petioles. Flowers are in large corymbs (fig. 386), each flower up to 5 cm across (fig. 387), and occur in profusion during December and January. Found along lowland to montane streamsides and forest margins from north-west Nelson to just south of Westport.   ASTERACEAE

388   Yellow eyebright, *Euphrasia cockayniana,* is a low, succulent herb, 5–10 cm high, which produces bright yellow flowers, 12–14 mm long, either singly or in pairs, towards the tips of the branches from December to March. Found in damp herbfields and boggy places from the Paparoa Range to Arthur's Pass.   SCROPHULARIACEAE

388   Yellow eyebright in flower, Arthur's Pass (January)

389–390   Tararua eyebright, *Euphrasia drucei,* is a perennial herb (fig. 389) with erect branches, 5 cm high, and close-set, sessile leaves, 2–10 mm long by 1–5 mm wide, having thickened margins rolled slightly outwards. Flowers, 10–15 mm across (fig. 390), occur in great abundance from December to February. This eyebright is found from the Ruahine Range south to Fiordland in damp or boggy places in alpine herbfields, fellfields and grasslands.

SCROPHULARIACEAE

389   Tararua eyebright in full flower, Mt Holdsworth (February)

390   Flowers of tararua eyebright, close up, Mt Holdsworth (February)

391   Creeping eyebright, *Euphrasia revoluta,* forms a creeping plant, up to 10 cm high, often found in subalpine and alpine damp herbfields, tussock grass-lands and on the edges of bogs from the Ruahine Range southwards. Large flowers, 10–15 mm across, occur on slender stalks from December to March.

SCROPHULARIACEAE

391   Creeping eyebright flowers, close up, Mt Holdsworth (December)

392 New Zealand eyebright, *Euphrasia cuneata,* is a herb up to 60 cm high, with smooth, wedge-shaped leaves, 5–15 mm long, with slightly thickened, flat margins. Flowers, 15–20 mm long, occur in profusion from January to March. Found in damp herbfields, fellfields, subalpine scrub, damp, rocky places and along streamsides up to 1,550 m altitude, throughout the North Island and as far south as North Canterbury. SCROPHULARIACEAE

392 New Zealand eyebright in flower, Mt Holdsworth (February)

393 Lesser New Zealand eyebright, *Euphrasia zelandica,* is a sparingly branched, rather succulent, hairy herb, 5–20 cm high. The sessile, hairy leaves, 4–10 mm long by 2–6 mm wide, have thickened, recurved margins and tend to cluster as rosettes at the branch tips. Flowers, about 10 mm long, occur during January and February. Found in damp places in herbfields and fellfields from Mt Hikurangi southwards. A similar but smaller eyebright, without hairs on the leaves, *E. australis,* is found throughout the Fiordland mountains. SCROPHULARIACEAE

393 Lesser New Zealand eyebright in flower on Mt Ruapehu (January)

394 Nelson eyebright growing among *Celmisia sessiliflora* on Mt Robert, Nelson (January)

394 Nelson eyebright, *Euphrasia townsonii,* is a tufted herb with slender stems rooting at the nodes and with erect branches about 15 cm high. The leaves, 4–15 mm long by 2–7 mm wide, are sessile; the flowers, 15–20 mm long, occur singly towards the tips of the branches from December to February or March. Found on the mountains of north-west Nelson and the Paparoa Range.

SCROPHULARIACEAE

395 New Zealand portulaca plant with flowers, Gertrude Cirque, Homer (January)

395 New Zealand portulaca, *Claytonia australasica,* is a succulent, spreading herb, forming patches up to 15 cm across in wet subalpine grasslands and herbfields and along streamsides from the Central Volcanic Plateau southwards. Flowers, 2 cm across, either solitary or paired, occur from November to January. PORTULACACEAE

396   Green cushion daisy plant in flower, Cupola Basin (January)

397   Crenate-leaved cotula, *Cotula potentillina*, forms a creeping herb, with pinnate leaves, 4–8 cm long, and with the pinnae irregularly crenate along their margins. Flowers, 8–10 mm across, occur from November to January, and the plant is found in damp situations on the Chatham Islands.

ASTERACEAE

397   Crenate-leaved cotula with flowers, Otari (November)

398–399   Makaka/marsh ribbonwood, *Plagianthus divaricatus*, is usually a much-branching, deciduous shrub, or a small tree up to 4m high, but sometimes it becomes prostrate. Leaves on juvenile plants are 3 cm long by 5 mm wide; on adult plants 15 mm by 2 mm. Flowers (fig. 398), 5 mm across, are axillary and solitary with a strong sweet scent and occur from September to November, and the fruit, about 5 mm across, matures from December to March. Found usually alongside salty swamps or damp, gravelly places in coastal regions throughout New Zealand.

MALVACEAE

396   Green cushion daisy, *Celmisia bellidioides*, is a creeping, branching, rooting daisy, forming mats up to 1 m across on wet rocks or gravel through which water is slowly flowing. It is found from 600 to 1,600 m in the South Island mountains. The fleshy, slightly spathulate leaves are 8–12 mm long by 3–4 mm wide, and the flowers, up to 2 cm across, occur during December and January.

ASTERACEAE

398   Makaka, close-up of flowers, Colville (October)

399   Makaka spray, showing leaves and berries, Colville (December)

400 Cardamine, *Cardamine debilis,* is one of the six cardamines found throughout New Zealand. The leaves are compound. The flowers, 4–5 mm across, occur from November to January, and the plant is found in damp situations from the coast to subalpine herbfields.   CRUCIFERAE

400 Cardamine in full flower, Temple Basin (January)

402 Flowers and flower-buds of leathery-leaved mountain hebe, The Wilderness (December)

402 Leathery-leaved mountain hebe, *Hebe odora,* is a shrub to 1.5 m high, found in damp places in herbfields and fellfields throughout the New Zealand mountains. The leaves, 10–30 mm long, are imbricate, stiff, concave, leathery and tend to be uniform. Flowers in lateral spikes occur from October to March.   SCROPHULARIACEAE

401 Mountain koromiko, *Hebe subalpina,* forms a densely branched shrub, up to 2 m high, with pubescent branchlets and narrow, pointed leaves, 2.5 cm long by 5–8 mm wide. Found in damp montane to subalpine regions of Westland and Canterbury. The simple lateral flowers occur in profusion during December and January, and each is about 2.5 cm long.   SCROPHULARIACEAE

401 Mountain koromiko in full flower, Arthur's Pass (December)

403 Hairy forget-me-not flowers and leaves, Key Summit (January)

403 Hairy forget-me-not, *Myosotis forsteri,* has very hairy leaves, flower-stalks and buds; the leaves, 15–40 mm long by 10–30 mm wide, are on stout hairy petioles. Flowers, 2–6 mm across, occur from October till April. Found along streamsides and in forest from the Urewera southwards.

BORAGINACEAE

404 Early winter orchids in flower, Hinakura (September)

405 Plants of early winter orchid, Hinakura (September)

404–405 Early winter orchid, *Acianthus reniformis* var. *oblonga*, is a small ground orchid that grows as small clumps on damp banks (fig. 404) and lowland grassland or in scrub and among moss throughout the North Island but only occasionally in the South Island. Flowers, 8 mm across, occur during July and August, and the basal sessile leaf is 10–40 mm long (fig. 405).   ORCHIDACEAE

406 Tufted grass cushion with fruiting capsules, Kaimanawa Mountains (November)

406–407 Tufted grass cushion, *Colobanthus apetalus*, is a loosely tufted cushion plant (fig. 406) with grass-like, pointed leaves, 10–25 mm long. Flowers up to 6 mm across occur during December and January, and the seed capsules mature during March and April, later they open to display round, black seeds (fig. 407).   CARYOPHYLLACEAE

407 Close-up of fruiting capsule of tufted grass cushion, Kaimanawa Mountains (November)

408–409 Dense sedge, *Uncinia uncinata*, forms clumps (fig. 408) up to 50 cm high in shaded, open places in lowland and subalpine forest up to 900 m altitude, throughout New Zealand. It is common in clearings and alongside walking tracks. Flowers occur from November to February as dense spikes, up to 15 cm long, which are shorter than the leaves (fig. 409).   CYPERACEAE

408   Plant of dense sedge, Tauhara Mountain (January)

410   Potts' forget-me-not, flower and leaf, Norman Potts garden, Opotiki (November)

410   Potts' forget-me-not, *Myosotis petiolata* var. *pottsiana*, is an open-branched plant with rosettes of leaves, 15–25 mm long by 10–17 mm wide, and flowers 9–12 mm across, occurring from November to February. Found along the Otara River near Opotiki.   BORAGINACEAE

409   Flower-heads of dense sedge, Tauhara Mountain (January)

411   Bush rice grass, *Microlaena avenacea*, is a grass, often quite abundant in damp, shaded places in lowland and subalpine forests throughout New Zealand. Plants reach a height of 1.3 m, and the graceful flower panicles appear during December and January.   GRAMINEAE

411   Bush rice grass flower panicles, Tauhara Mountain (January)

412   Scrambling broom, stem with
typical seeds, Otari (February)

412   Scrambling broom, *Carmichaelia (Kirkiella)*
kirkii, is a semi-prostrate plant, which scrambles
over the ground with long, thin, interlacing stems
that bear leaves only during spring and summer.
Flowers occur as open racemes followed by the
stout, straight-beaked seed-pods shown in fig. 412.
Found on lowland wet river terraces and streamsides from Canterbury southwards.   FABACEAE

416   Flowers and leaves of Traver's hebe, French Pass
(January)

416   Traver's hebe, *Hebe traversii*, is a shrub, up
to 2 m high, with slender branches and narrow,
spreading leaves, about 2.5 cm long by 4–7 mm
wide, found along damp streamsides and banks in
montane to subalpine regions from Marlborough
to mid-Canterbury. Flowers arise laterally from
December to March.   SCROPHULARIACEAE

413–415   Koru, *Pratia physalloides*, is an erect-
branching herb, 1 m high, woody towards its base
and with large, soft, conspicuously veined and
toothed leaves, up to 20 cm long by 10 cm wide (fig.
414). The odd-shaped flowers (fig. 413) on hairy
peduncles occur as terminal racemes from October
till May, and the berries, 10–15 mm long, are ripe
from April on (fig. 415). Found in shaded places
along coastal and lowland forest margins and
streamsides in the northern part of the North Island
and the adjacent offshore islands.   LOBELIACEAE

413   Koru flowers, Waipahihi Botanical Reserve
(February)

414   The striking leaves of koru, Waipahihi Botanical
Reserve (February)

415   Spray of koru with berries, Waipahihi Botanical
Reserve (May)

PLANTS OF FORESTS

The forests of New Zealand are evergreen, made up of massive timber trees forming the upper or primary canopy and, below them, lesser trees, which form secondary canopies. Below these again is a thick undergrowth of shrubs, lianes, ferns, mosses, lichens, liverworts and numerous epiphytic and parasitic plants. Many of these trees produce spectacular flowers and fruits, and this, along with the association within the forest of many and diverse species, imparts a richness and extravagance to the forest that is more typical of subtropical than of temperate regions. In places, such as on Mt Taranaki and the Westland forests, this assemblage of plants is a typical rain forest, and in its primeval condition formed a dense, in places almost impenetrable, barrier extending from sea-level to 1,200 m in the warmer north but ascending to 900 m in the cooler south. Known to New Zealand people as 'the bush' these mixed forests were dominated by different species in different localities, and few extensive stands of any single species, other than beech, were known. Since European settlement of the country began, most of these forests have disappeared and, today, the original primeval New Zealand forests can only be seen in some National Parks and special reserves such as those at Pureora, Whirinaki and South Westland.

Forest trees, like other plants, have preferences for differing habitats; some prefer growing where they have cool, damp root runs, others prefer drier places, while others prefer warmer or cooler climates and yet others thrive near the sea while their opposites thrive high in the mountains. Some grow best on limestone-based soils while others do not. Steep, shaded banks, bogs and swamps can occur within forests as well as in open country and, accordingly, as we move into the forest arena, we look first at those trees that prefer damp or wet situations in which to grow.

DAMP AND WET PLACES

417   Kahikatea forest exposure, Lake Rotoroa (April)

417–420   Kahikatea/white pine, *Dacrycarpus dacry-dioides*, is New Zealand's tallest tree; specimens growing in the estuary of the Kauaeranga River, near what is now Thames, were measured by Cook, on the *Endeavour*, at over 200 feet high. Some small stands of such trees still exist in South Westland. Kahikatea prefers growing in wet swamps but can also grow on dry land and even on dry hillsides, and is found naturally in wet and damp areas throughout New Zealand forests from sea-level to 600 m. Fig. 417 shows a stand of kahikatea on the shore of Lake Rotoroa; fig. 418 shows kahikatea foliage and the black seeds in their red receptacles, while fig. 419 shows the seed and receptacle close up. Fig. 420 shows the ripe male cones of kahikatea.

PODOCARPACEAE

418   Mature seeds of kahikatea, Lake Pounui (April)

419 Close-up of a mature kahikatea seed, showing the mature bluish-coloured seed on top of the orange-red-coloured receptacle, Lake Pounui (April)

420 Mature male cones of kahikatea, Lake Pounui (October)

421–422 **Kawaka,** *Libocedrus plumosa*, is a tree up to 25 m high with connate, sheathing, compressed leaves, 2.5–5 mm long, arranged in four rows on the branchlets (fig. 421). It is found in damp lowland forests from Mangonui south to about Opotiki, and from Collingwood to Westhaven. Male cones (figs 421–422) are produced sparingly during September and October.   CUPRESSACEAE

421 Spray of kawaka showing male cones, Opotiki (September)

422 Close-up of male cones of kawaka, Opotiki (September)

423–426 Pahautea/mountain cedar, *Libocedrus bidwillii,* is a tree up to 20 m high found in damp places in montane and subalpine forests from the Coromandel Peninsula south to the forests of Fiordland. The appressed, triangular, pointed leaves are 2 mm long, and the male cones appear in profusion (fig. 423) during September and October (fig. 424). Female cones (fig. 426) appear in October, and seeds (fig. 425) are ripe by November.

CUPRESSACEAE

423 Spray of pahautea bearing mature male cones, Tongariro National Park (September)

424 Close-up of male cones of pahautea, Tongariro National Park (September)

426 Pahautea showing leaves and newly formed female cones, Otari (October)

425 Close-up of a ripe seed of pahautea, Otari (November). Kawaka has a similar seed.

427 Maire tawaki, ripe berries, Lake Pounui (October)

427–429 Maire tawaki, *Syzygium maire*, forms a tree up to 15 m high, with a smooth, white bark, four-sided branchlets and sinuate, opposite leaves, 4–5 cm long by 10–15 mm wide, on petioles 5–10 mm long (fig. 428). Flowers (figs 428–429), about 12 mm across, occur from January to March, and the berries (fig. 427) ripen from March to July. Found throughout both islands in lowland to montane, boggy and swampy forests.   MYRTACEAE

429 Close-up of flower of maire tawaki, Lake Pounui (March)

428 Opening flower-head of maire tawaki, Lake Pounui (March)

431 Flowers of mountain lacebark, *Hoheria glabrata*, Rahu Saddle (January)

430 Flowers of mountain lacebark, *Hoheria lyallii*, Peel Forest (January)

430–431 Mountain lacebarks, *Hoheria lyallii* (fig. 430) and *Hoheria glabrata* (fig. 431), are both deciduous shrubs or small trees, 6–10 m high, found in wetter places in subalpine forest and scrub in the South Island, *lyallii* on the east side, *glabrata* on the west side of the Southern Alps. The thin, delicate leaves, 2–7 cm long by 2–6 cm wide in *lyallii*, are usually smaller in *glabrata*, and the branchlets, leaves and stems are more heavily clothed with stellate hairs in *lyallii*. Flowers, 2–4 cm across, occur in profusion from November to February.   MALVACEAE

432 Hunangamoho with flowers, Hollyford Valley (January)

432–433 Hunangamoho, *Chionochloa conspicua*, is a large tussock, up to 2 m high, found throughout New Zealand in lowland and subalpine forests and natural forest clearings, especially near streams and seepages. The leaves are strongly nerved, 45 cm–1.2 m long by 6–8 mm wide, flat and often hairy along the margins. Handsome flower panicles, about 45 cm high, occur from November to January.
                                        GRAMINEAE

433 Flower panicle of hunangamoho, Hollyford Valley (January)

434–438 Pukatea, *Laurelia novae-zelandiae*, is a large, aromatic tree, up to 35 m high, with a 2 m diameter trunk buttressed around its base. Leaves are glossy, serrated, thick and leathery, 4–8 cm long by 2.5–5 cm wide, on petioles 10 mm long. Flowers are unusual, occurring as perfect, with both male and female parts in the one flower (fig. 436); male flowers (fig. 434) have stamens (red) and female flowers have stigmas but no stamens (fig. 435); all as well have yellow, triangulate structures called staminodes. Seeds (figs 437–438) are contained in elongated, pear-shaped seed-cases. Pukatea is found in lowland swampy forest, creek beds and damp gullies throughout the North Island and south to Fiordland. MONIMIACEAE

434 Male flowers of pukatea, Kaitoke (November)

435 Close-up of female pukatea flowers, Kaitoke (November)

437 Pukatea seed-cases opened releasing wind-borne seeds, Waikanae (May)

436 Close-up of perfect pukatea male flower, Kaitoke (November)

438 Pukatea spray showing unopened and opened seed-cases, Waikanae (May)

439–440 Cabbage tree/ti kouka, *Cordyline australis*, grows to a height of 12–20 m with an unbranched trunk topped by a mass of leaves and flowers. Cabbage tree is a collective name for five species, of which ti kouka is the most common, being found in damp situations along forest margins and open spaces, especially alongside bogs and swamps, throughout the country from sea-level to 600 m altitude. Isolated specimens often occur on hillsides near seepages. The long leaves are are up to 1 m long by 3–6 cm wide. Sweet-scented flowers, each 2 cm across, arise on dense panicles up to 1.5 m long, above the leaf canopy, from September to December followed in January–February by whitish berries speckled with blue.   ASPHODELACEAE

441 Toi tree with flowers, Mt Taranaki (December)

439 Cabbage trees heavy with flowers, Hinakura (November)

441–445 Toi/broad-leaved cabbage tree/mountain cabbage tree, *Cordyline indivisa,* is similar to ti kouka but has longer, broader leaves, 1–2 m long by 10–15 mm wide, and grows only in the wetter mountains from the Hunua and Coromandel Ranges to Fiordland. Sweet-scented flowers occur from November to January as tightly packed panicles (fig. 445) in a flower-head up to 1.6 m long by 30 cm wide, which arises beneath the leaf canopy (fig. 442), the individual flowers up to 2 cm across (fig. 444). This flower-head is followed by a mass of black berries, each about 6 mm in diameter, during April and May (fig. 443).   ASPHODELACEAE

440 Close-up of cabbage tree flowers, Lake Pounui (November)

446 Dwarf cabbage tree, *Cordyline pumilio*, seldom reaches 2 m high and is sometimes stemless, when it can be mistaken for a tussock or a sedge. The stiff leaves are up to 60 cm long by 10–20 mm wide. Small, very sweet-scented flowers, 5 mm across, arise, widely spaced, along the panicle, which is up to 60 cm long, during December and January. Found in the North Island only in damp places in open lowland forest and scrub. ASPHODELACEAE

442 The pendant flower-head of toi, Mt Taranaki (December)

443 The pendant fruiting head of ripe berries of toi, Mt Taranaki (May)

444 Close-up of an individual toi flower, Mt Taranaki (December)

446 Dwarf cabbage tree plant, Waipahihi Botanical Reserve (October)

445 A section of flower-head of toi, Mt Taranaki (December)

447–450 Yellow pine, *Halocarpus biformis*, in forest forms a small tree, 10 m high, but on exposed mountainsides it forms a tight, rounded, cypress-like shrub or small tree, sometimes only 1 m high. It occurs from the central Volcanic Plateau south to Stewart Island, from sea-level to 1,400 m, usually in damp places. Adult leaves are 2 mm long, densely imbricate, appressed and keeled; juvenile leaves and leaves of reversion shoots are 10–20 mm long and spreading (fig. 450). Male cones (fig. 448) are produced in profusion during October, shedding pollen during January (fig. 447). Small female cones appear in December and, after pollination, grow slowly to reach maturity fourteen months later (fig. 449).

PODOCARPACEAE

447　Yellow pine, foliage with opened male cones, Mt Ruapehu (February)

451　Yellow silver pine, young and adult foliage, Taita (October)

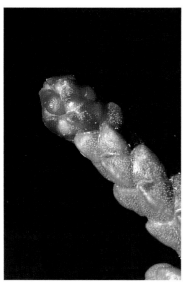

448　Yellow pine, mature male cone, Mt Ruapehu (December)

449　Yellow pine, mature seed sitting on yellow aril, Mt Ruapehu (February)

450　Juvenile foliage above with adult foliage below of yellow pine, Mt Ruapehu (December)

452 Mature female cone of yellow silver pine, Coromandel Peninsula, (April)

453 Male cones of yellow silver pine, Taita (October)

454 Close-up of male cones of yellow silver pine, Taita (October)

451–454 Yellow silver pine, *Lepidothamnus intermedius*, is a spreading tree up to 15 m high, found in wet places in lowland, montane and subalpine forests from the Coromandel south to Stewart Island; in the latter it is the principal tree in swamp forests. The imbricate adult leaves are 1.5–3 mm long; juvenile leaves are 9–15 mm long and spreading (fig. 451). Male cones (fig. 454) mature in profusion from October to December (fig. 453), and female cones are mature by April (fig. 452).

PODOCARPACEAE

455–456 Silver pine, *Lagarostrobos colensoi*, is a somewhat spreading tree, up to 15 m high, found from Mangonui to Mt Ruapehu and, in the South Island, throughout Westland in shady places with a rich soil and a wet climate. Leaves of juveniles, 6–12 mm long, are spreading; leaves of adults, 1–2.5 mm long, are imbricate, appressed and keeled. Male cones (fig. 455) occur abundantly from September to November; female cones are borne sparingly and take about eighteen months to mature (fig. 456), usually in November, but at irregular intervals.

PODOCARPACEAE

455 Silver pine, male cone sheding pollen, Mt Ruapehu (November)

456 Mature female cone of silver pine, Mt Ruapehu (March)

457–459 Kotukutuku/tree fuchsia/konini, *Fuchsia excorticata*, forms a deciduous shrub or a small tree, up to 12 m high, with a red, papery bark that peels readily to reveal a smooth, satiny, greenish inner layer. Leaves, 3–10 cm long by 15–30 mm wide, are pale green or pale silvery below on petioles 10–40 mm long. Flowers, notable for their blue-coloured pollen, begin to arise in August, in some districts, and become abundant during October and November, often persisting through to February. Berries (fig. 459) occur from September to February. Found throughout New Zealand and the Auckland Islands in lowland and montane forests, especially along forest margins and streamsides where the soil is damp.   ONAGRACEAE

458 A kotukutuku flower, close up, Wellington (October)

457 Kotukutuku branch with flowers displaying typical blue-coloured pollen, Lewis Pass (January)

460 Sprawling fuchsia, *Fuchsia colensoi*, is a low, sprawling, woody shrub with long straggling branches bearing variable-shaped leaves and numerous flowers very similar to those of *F. excorticata*. The berries, however, are narrower and more cylindrical in shape than those of *F. excorticata*.

ONAGRACEAE

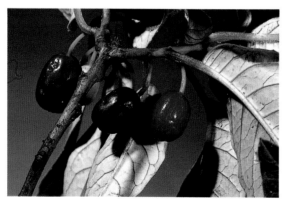

459 Kotukutuku spray with berries, Kau Kau, Wellington (January)

460 Sprawling fuchsia with flowers and berries, Waikanae (December)

461-462 Round-leaved coprosma, *Coprosma rotundifolia*, is a slender shrub or small tree, up to 5 m high, with rounded leaves (fig. 461), up to 25 mm long, and entangled branches. It is found throughout New Zealand in damp lowland and montane forests or in scrub and along riverbanks. Slender, drooping flowers (fig. 462) arise during September and October, and the drupes (fig. 461), which ripen during February and March, occur singly or in twos and threes along the branches.

RUBIACEAE

463-464 Great Barrier tree daisy, *Olearia allomii*, is a shrub about 1 m high with very thick, leathery leaves, 2.5-5 cm long (fig. 464). It is found only on the Great Barrier Island, in damp forest from sea-level to 700 m. Sweet-scented flowers (fig. 463), 15 mm across, occur during October and November.

ASTERACEAE

463 Great Barrier tree daisy, close-up of flowers, Otari (October)

461 Round-leaved coprosma drupe and leaves, Pelorus Bridge (January)

464 Great Barrier tree daisy, spray with flowers and leaves, Otari (October)

462 Branchlet of round-leaved coprosma with male flowers, Silverstream (October)

COASTAL FOREST

Several trees grow in coastal areas, mostly as isolated groves or specimens, but in its primeval state New Zealand's coasts, in many places, were clothed by coastal forests. Today, these are represented only by scattered reserves such as that near Paraparaumu, which preserves a remnant of the once-dominant coastal kohekohe forests.

465   Panicles of flowers sprouting from the trunk of a kohekohe tree, Huntleigh Park, Wellington (May)

465–468   Kohekohe, *Dysoxylum spectabile*, is a tree, up to 15 m high, that grows in damp areas of lowland coastal forests from North Cape to Nelson and once formed extensive coastal forests in restricted areas. From March till June the tree produces spectacular, long, drooping panicles of greenish white, waxy flowers (figs 466–467), sprouting directly from the trunk and branches (fig. 465). These are followed from April to September by panicles of green capsules, 2.5 cm across, which split to reveal bright orange or red arils (fig. 468) that carry the seeds.

MELIACEAE

466   Portion of a flower panicle of a kohekohe tree, Huntleigh Park, Wellington (May)

468   Seed capsules of kohekohe opening to reveal the red arils, Waikanae (May)

467   Close-up of a kohekohe flower, Waikanae (June)

469–472 Nikau, *Rhopalostylis sapida,* is a palm reaching to 10 m in height, found in lowland and coastal forests throughout the North Island and, in the South Island, as far south as Banks Peninsula and Greymouth. The rings on the trunk are leaf scars from feather-like leaves up to 3 m long and 2 m wide. The flower panicle shown in fig. 469 has the upper bud sheath behind it; this usually falls within minutes of the opening of the flower. To the left in fig. 469 are the green berries from the previous year's flower, and below are the brownish seeds from two years before. Flower-spikes are about 30 cm long and arise from December to February. Fig. 471 shows male flowers closely packed on the flower-spike, with occasional tiny female buds between them. There are three male flowers to each single female, and fig. 470 shows a male flower, close up, with the tiny female flower beside it to the right and another female to the left, beneath the tips of the two anthers. Female flowers do not fully open until after the male flowers have fallen off. The orange to red berries are ripe twelve months later (fig. 472).

PALMAE

470   Close-up of nikau flowers: male flowers with anthers, female flowers are the tiny bud-like structures beside the two male flowers, Hukutaia Domain (December)

471   A section of a panicle of nikau flowers, females visible between male flowers, Hukutaia Domain (December)

469   The flower panicle (above), last year's seeds (middle) and the year before's seed remain on a nikau palm, Lake Pounui (February). Note the flower-sheath behind the flower panicle.

472   Ripe nikau berries, Waiorongomai (April)

473 Ti ngahere trees, Wharariki Beach (January)

475 Young ti ngahere tree in flower, Kauaeranga Valley (December)

473–477 Ti ngahere/forest cabbage tree, *Cordyline banksii*, unlike other cabbage trees, usually has several stems, up to 4 m high (fig. 473), and tends to be more bushy. It grows, preferably in damp situations, in lowland to montane forest, in coastal forests, or along forest margins from North Cape to North Westland. Fragrant white flowers in rather open panicles (fig. 474), 1–2 m long, occur from November to January, and whitish berries with blue markings (fig. 477) follow in February, persisting till April. Quite small plants, only 1 m high, will flower (fig. 475). AGAVACEAE

476 Close-up of flower of ti ngahere tree, Kauaeranga Valley (December)

474 Flower panicle of ti ngahere tree, Lake Pounui (November)

477 Ti ngahere berries, Lake Pounui (February)

478 Akeake seeds, Ure River Valley (March)

479 Spray of akeake with male flowers, Otari (October)

480 Close-up of male flowers of akeake, Otari (October)

478–482 **Akeake,** *Dodonaea viscosa*, is a shrub or small tree, up to 6 m high, with narrow leaves up to 10 cm long (figs 478–479); the branches and branchlets are usually sticky. The very small male flowers (fig. 479) occur from September to January, and the unusual-looking fruits (fig. 478) from November to March. It is found from North Cape to Banks Peninsula in the east and Greymouth in the west. The thin, reddish-coloured bark peels in thin flakes, and the sexes occur on separate trees (fig. 480, male flowers; fig. 481, female flowers). A variety, *purpurea*, with copper-coloured leaves and copper-coloured fruits (fig. 482), is known and is often grown in gardens. SAPINDACEAE

482 Seeds of the purple form of akeake, Taupo (December)

481 Female flowers of akeake, Otari (October)

LOWLAND AND MONTANE FORESTS

Lowland forests are inland from coastal forests and usually situated on rolling or flat country. They are rich in species and included much of the original primeval podocarp dominant forests of New Zealand. As the land rises and slopes become steeper, we move into montane forests of mixed podocarp and other species, which extend towards the upper subalpine and alpine forests that consist, in the main, of beech trees.

483–484 Coastal tree daisy, *Olearia albida*, is a shrub up to 5 m high, with grooved branchlets clothed with a loose, white tomentum. The leathery leaves, 7–10 cm long by 2.5–3.5 cm wide, are on petioles 2 cm long (fig. 483). Flowers (fig. 484) appear from January to May, and the plant is found in coastal forests from North Cape to East Cape and Kawhia.   ASTERACEAE

483   Coastal tree daisy, spray with flower-heads, Seatoun, Wellington (April)

485   Kohuhu spray with flowers, Karaka Bay (September)

485 Kohuhu, *Pittosporum tenuifolium*, is a small tree, up to 8 m high, found commonly throughout New Zealand, except in Westland, in lowland and coastal forests. The thin, glossy, wavy leaves are 3–7 cm long by 10–20 mm wide. Flowers, 12 mm across, are produced in abundance from September to November and, in the evening, fill the air with a strong, sweet fragrance.   PITTOSPORACEAE

484   Close-up of flowers of coastal tree daisy, Seatoun, Wellington (April)

486–488   Kauri, *Agathis australis*, is New Zealand's most massive tree, reaching a height of 30–60 m with a trunk diameter of 3–7 m. Found in lowland forests in Northland, the Coromandel and south to Maketu in the east and Kawhia in the west. Male and female cones occur on the same tree (fig. 486); the male (fig. 487) appears during August and September, the female (fig. 488) during September and October. The thick, parallel-veined adult leaves are 2–3.5 cm long, those of juvenile trees are 5–10 cm long by 5–12 mm wide. The male cone is mature and releases pollen when one year old; the female opens for pollination, also when about twelve months old, but does not open to release seeds until another two years have passed.   ARAUCARIACEAE

487   Kauri branchlet with mature male cone, Waipoua (November)

486   Kauri branch with two immature male cones and one female, Waipoua (October)

488   Mature female kauri cones, Waipoua (September)

489   Tarairi flowers and leaves,
       Whangarei (December)

491   Tarairi drupes, Whangarei
       (May)

489–491   Tarairi, *Beilschmiedia tarairi*, forms a medium-sized tree, up to 22 m high, found in lowland forests, especially kauri forests, in Northland and as far south as Kawhia and Tauranga. The prominently veined, glossy, leathery leaves, 4–15 cm long by 3–6 cm wide, on petioles 10–15 mm long, have the veins below and the petioles clothed with a thick brown to golden-coloured wool (fig. 489). Small apetalous flowers (fig. 490), each 5 mm across, occur in panicles up to 10 cm long from September to December (fig. 489).The purple-bloomed drupes (fig. 491), up to 3.5 cm long, ripen during the following April and May.   LAURACEAE

492   Branch of wavy-leaved
       coprosma heavy with drupes,
       Opepe Bush (May)

492   Wavy-leaved coprosma, *Coprosma tenuifolia*, is a shrub or small tree to 5 m high, with wavy-margined leaves, 4–10 cm long. The drupes, 7–8 mm long, ripen in April and May. The plant occurs in lowland to montane forests and in subalpine scrub from Te Aroha Mountain south to the Ruahine Range; it is very common on Mt Taranaki.

RUBIACEAE

490   Tarairi flowers, close up, Colville (December)

493 Kanono/raurekau, *Coprosma australis,* is a shrub or small tree to 6 m high, found in lowland and montane forests and along forest margins or in scrub throughout the country. The leaves, 10–20 cm long by 5–10 cm wide, are on petioles 2–5 cm long. Flowers occur from April to June, often along with the drupes of the previous season, which are 9 mm long and take some twelve months to ripen.

RUBIACEAE

494–496 Tawa, *Beilschmiedia tawa,* is a tree to 25 m high, with leaves 5–10 cm long by 10–20 mm wide on petioles 10 mm long. Panicles, about 8 cm long, arising from the axils of the leaves and carrying tiny flowers (fig. 494), 2–3 mm across (fig. 495), appear from September to December. The drupe, 2–3 cm long (fig. 496), appears from October to February. Tawa is found in lowland and montane forests from North Cape to Nelson and Marlborough.

LAURACEAE

493 Kanono showing leaves and drupes, Opepe Bush (May)

494 Tawa spray with flower panicles, Barton's Bush, Silverstream (December)

495 Tawa flower, Barton's Bush, Silverstream (December)

496 Tawa drupe, Wilton's Bush, Wellington (March)

497–499 Makamaka, *Ackama rosaefolia,* is a small
bushy tree, up to 12 m high, with pinnate, serrated
leaves having a terminal pinna (fig. 497). Found in
lowland forest and along forest margins and streamsides from Mangonui to about Dargaville. The
flowers arise in much-branched panicles (fig. 497),
up to 15 cm long, with each flower 3 mm across (fig.
498), from August till November, and the fruits (fig.
499) ripen from January to March.

CUNONIACEAE

497   Makamaka flowers and leaves, Otari (October)

499   Makamaka berries, Otari (April)

500   Putaputaweta flowers, close up, Lake Pounui
(December)

498   Close-up of makamaka
flowers, Otari (October)

500–501 Putaputaweta, *Carpodetus serratus,* is
usually a small tree up to 10 m high, with flat, rather
fan-like branches bearing leaves 4–6 cm long (fig.
501). White, star-like flowers, 5–6 mm across (fig.
500), occur as broad panicles (fig. 501) from
November to March. The fruit is a black capsule,
4–6 mm across, that ripens from March through
May. Putaputaweta is found throughout New
Zealand in lowland and montane forests and along
forest margins and streamsides and in natural
clearings.   ESCALLONIACEAE

501   Spray of putaputaweta
with flowers and leaves,
Lake Pounui (December)

502 Rimu foliage with ripe female cones, Mangamuka (February)

503 Spray of rimu with male cones, Bushy Park (November)

502–503 Rimu/red pine, *Dacrydium cupressinum*, is a tall forest tree, up to 60 m high, found in lowland and montane forests throughout the North, South and Stewart Islands. With its characteristic feather-like foliage (fig. 502), rimu is one of New Zealand's finest timber trees. Female cones with their red arils (fig. 502) appear at irregular intervals over several years, usually about April through May. The same applies to the male cones (fig. 503), which appear late November to December in a year when they occur. PODOCARPACEAE

505 Hutu branch with flowers, Otari (September)

504 Hutu leaves, Hukutaia Domain (June)

504–505 Hutu, *Ascarina lucida*, is a closely branched, small, highly aromatic tree, up to 8 m high, found in lowland and montane forests from Hokianga Harbour southwards to Stewart Island but more common in the South Island. The branchlets are purple coloured (fig. 504), the leaves, 2–7 cm long by 15–35 mm wide, are serrated (fig. 504) and on petioles about 10 mm long. Flowers as spikes occur from September to November (fig. 505), followed by small, whitish berries from October to January. CHLORANTHACEAE

506   Pigeonwood spray with male
flowers, Waiorongomai (October)

507   Close-up of male pigeonwood flower,
Waiorongomai (November)

508   Female pigeonwood flowers, Lake Pounui
(December)

506–509 Pigeonwood/porokaiwhiri, *Hedycarya arborea*, is an aromatic tree, up to 12 m high, with thick, tough, wavy-margined leaves, 5–12 cm long by 2.5–5 cm wide, on petioles 2 cm long. Racemes of lemon-scented flowers (fig. 506), each flower about 10 mm across, occur from September to December. The sexes are on separate trees, with the male (fig. 507) and the female (fig. 508) quite different. The drupes (fig. 509) take twelve months to mature and are ripe from October to March of the following year. Pigeonwood is found in damp situations in lowland and montane forests from North Cape to Banks Peninsula.   MONIMIACEAE

509   Pigeonwood drupes, Otaki Forks (March)

510 Mingimingi, *Coprosma propinqua*, is a small-leaved shrub or small tree, 3–6 m high, found throughout New Zealand in lowland forest, along forest margins and streambanks, in scrub, gravelly places and along the edges of bogs and swamps. Flowers occur during October and November, and the white or blue drupes are mature from March to May.   RUBIACEAE

510   Mingimingi branchlets with leaves and drupes,
Ure River Gorge (March)

511 Titoki spray showing leaves
and flowers, Barton's Bush
(November)

512 Winged seed capsules of
titoki opening to show black
seeds sitting on red arils,
Barton's Bush (December)

513 Close-up of titoki flower, Barton's Bush
(November)

511–513 **Titoki,** *Alectryon excelsus,* is a spreading
tree, up to 10 m high, with pinnate leaves (fig. 511),
10–40 cm long. It is found in lowland forests,
growing on alluvial soils from near North Cape south
to Banks Peninsula. Apetalous flowers (fig. 513)
arise as panicles (fig. 511) during October and
November, but the seeds that follow (fig. 512) take
one year to ripen so that flowers and seeds can occur
together on the tree.   SAPINDACEAE

514 *Coprosma rubra* branchlet
with leaves and drupe, Otari
(September)

514 *Coprosma rubra* is a shrub about 2–3 m high
with stems and leaf petioles clad with soft hairs and
the leaves, 10–15 mm long by 10–12 mm wide, are
strongly reticulated with ciliated margins when
young. Fig. 514 shows the elongate yellow drupe,
which occurs singly, but sometimes in groups of two,
and the female flower. This *Coprosma* grows in
lowland forest and scrub from the Pahaoa River in
the Wairarapa south to about Banks Peninsula.
                                          RUBIACEAE

515–517 Matai/black pine, *Prumnopitys taxifolia,* forms a robust forest tree, up to 25 m high, with a broad crown and narrow, elongate leaves, 5–10 mm long by 1–2 mm wide. Male cones (figs 516–517) arise on spikes about 5 cm long during October and November, and the fruit ripens during the following year to a black drupaceous seed with a purplish bloom (fig. 515). Matai is a common tree in lowland forests of both the North and South Islands but is rare on Stewart Island.   PODOCARPACEAE

515   Matai seeds, Hinakura (February)

517   Close-up of mature matai male cones, Lake Pounui (November)

516   Male cones of Matai, Lake Pounui (November)

518 Miro/brown pine, *Prumnopitys ferruginea,* forms a round-headed tree up to 25 m high, with sickle-shaped leaves, 10–25 mm long by 1–2 mm wide, that are in one plane. Male cones similar to those of *P. taxifolia* occur during October and November, and the brilliant seeds (fig. 518) mature one year later. Miro is a common tree in lowland forests throughout New Zealand.

PODOCARPACEAE

518   Miro seeds, Huntleigh Park, Wellington (April)

519 Totara spray with cones, Hinakura (April)

520 Immature totara seed on its aril, Woodside Gorge (September)

519–521 Totara, *Podocarpus totara*, is a tall forest tree, up to 30 m high, with thick, stringy, furrowed bark and narrow, leathery, pungent leaves, up to 3 cm long by 3–4 mm wide. Male cones (fig. 521) are abundant during October and November, each 15 mm long, and produced singly or four together on a short peduncle. The seeds sit in red, fleshy arils (fig. 520). Totara grows throughout New Zealand in lowland and montane forests, often as small pure stands. PODOCARPACEAE

521 Ripe male cones of totara, Akatarawa Saddle (November)

522 Montane totara/Hall's totara, *Podocarpus cunninghamii*, is a similar but smaller tree than *P. totara*, but differs in having thin, papery bark and subsessile leaves. Male cones and seeds are similar, with the receptacle in *P. cunninghamii* tending to be more elongate. Montane totara is found in lowland, montane and subalpine forests throughout New Zealand. PODOCARPACEAE

522 Ripe female cone of montane totara, Moawhango (April)

523–526 Five finger/pouahou, *Pseudopanax arboreum*, is a much-branching shrub or small tree, up to 8 m high, with compound leaves, each of 5–7 serrate-dentate-margined leaflets, 10–15 cm long (fig. 523). Five finger flowers profusely from June to August (fig. 523), the flowers in umbels, each flower 6 mm across (fig. 524). Sexes are on separate trees (figs 524–525). The rounded flower-buds are shown in fig. 526 but the dark brown to black seeds that occur from August to February are noticeably flattened. Five finger is common in lowland and montane forests and lowland scrub throughout New Zealand. ARALIACEAE

525 Close-up of female flowers of five finger, Taupo (August)

523 Five finger plant in flower, Taupo (August)

526 Flower-buds of five finger are easily confused with the dark, flattened, similarly arranged seeds that occur from September onwards, Karaka Bay (April)

524 Close-up of male flower of five finger, Taupo (June)

527 Haumakaroa, *Pseudopanax simplex*, forms a shrub or low, open-branched tree, 8 m high, in lowland forests from the Coromandel Ranges south to Stewart Island and on the Auckland Islands. Seedling plants have divided leaves that quickly change to 3–5 foliolate, then to the acuminate, sharply serrate, leathery, trifoliate and unifoliolate leaves, 5–10 cm long, with petioles 3–8 cm long, of adult plants. Umbels of 5–15 greenish-coloured flowers arise from June through to March. ARALIACEAE

527 Haumakaroa male flowers, close up, Taurewa Intake (January)

528–530 Lancewood/horoeka, *Pseudopanax crassifolium*, is a small, round-headed tree, 15 m high, with very thick, long, narrow, serrated leaves (fig. 530). The tree passes through distinct seedling and juvenile stages that bear little resemblance to the adult stage. Lancewood occurs throughout the lowland and montane forests of New Zealand, and in lowland scrub. Flowers arise in large terminal umbels (fig. 528) from January to April, and the fruits, large bunches of berries (fig. 529), each 5 mm across, ripen during the following twelve months.

ARALIACEAE

530   Spray of lancewood, showing leaf undersides and green berries, Lake Pounui (March)

528   Umbel of lancewood flowers, Waikanae (February)

529   Lancewood with ripe seeds, Lake Pounui (March)

531   Raukawa spray showing leaves, flowers, flower-buds and immature seeds, Mt Kaitarakihi (January)

531–532 Raukawa, *Pseudopanax edgerleyi*, is a tree 10 m high, with aromatic, green, shining, 1–3-foliolate leaves, paler below, and each 7–15 cm long by 3–5 cm wide. Greenish-coloured flowers (fig. 532) in small, 10–15-flowered, axillary umbels (fig. 531) arise from November to March; both the buds and seeds are rounded (fig. 532). Raukawa is found in lowland forests from about Mangonui southwards to Stewart Island.   ARALIACEAE

532   Close-up of flower, flower-bud and seed of raukawa, Mt Kaitarakihi (January)

533   Mapau showing ripe berries and leaves, Karaka Bay (April)

533–535   Mapau, *Myrsine australis*, is a shrub or small tree to 6 m high, found in lowland forests throughout New Zealand. The wavy-margined, leathery leaves (fig. 533) are 3–6 cm long by 15–25 mm wide and are on strong petioles 5 mm long. Flowers (male, fig. 534; female fig. 535) are 1.5–2.5 mm across and arise along the stems from December to April, followed in October through to February by clusters of berries (fig. 533).

MYRSINACEAE

535   Spray of mapau with female flowers, Whangarei (January)

534   Mapau male flowers shedding pollen, Karaka Bay (December)

536–537   Tawari, *Ixerba brexioides*, forms a canopy tree 6–16 m high, found from Mangonui to Mamaku, which smothers itself with creamy white flowers (fig. 536) in November and December. The narrow, leathery, serrated leaves (fig. 537), 6–16 cm long by 10–40 mm wide, are on petioles 2 mm long. Figure 537 shows a panicle of flower-buds, a seed capsule, and the unusual seeds revealed when the capsules open from January to April.   ESCALLIONIACEAE

536   The striking flowers of tawari set among terminal leaves, Rotorua (December)

538-539 Rewarewa/New Zealand honeysuckle, *Knightia excelsa*, is an upright branching tree, 30 m high, with long, thick, leathery, coarsely serrated leaves, 10-20 cm long by 2.5-4 cm wide. Brilliant flowers, in racemes up to 10 cm long (fig. 538), arise on the branchlets from October to December, and the seeds (fig. 539) mature by the following June. Rewarewa is found in lowland forests from North Cape to the Marlborough Sounds.   PROTEACEAE

538   Rewarewa flower, Ngaio Gorge, Wellington (November)

539   Rewarewa seeds, Otari (June)

540 Supplejack/karewao/pirita, *Ripogonum scandens*, is often met with as a mass of twining, twisting stems blocking progress through the forest. Supplejack is a climbing liane that, up in the light of the forest canopy, produces leaves, 5-16 cm long by 2-6 cm wide, and tiny flowers on branching stalks followed by clusters of berry-like fruits (fig. 540).

SMILACACEAE

537   Tawari spray showing seeds of this season and the flower-buds for next season, Mamaku Forest (May)

540   Supplejack spray with leaves and berries, Lake Pounui (October)

541 Flowers and leaves of toru,
Otari (October)

542 Close-up of toru flowers,
Otari (October)

541–543 Toru, *Toronia toru*, forms an erect tree,
12 m high, with narrow, alternate or whorled, thick,
leathery leaves (fig. 541), 16–20 cm long by 8–15 mm
wide. Fragrant flowers in 6–15-flowered axillary
racemes (figs 541–542) occur from October to
February and are followed by fruits clustered along
the stems from December to March (fig. 543). Toru
is found in lowland and montane forests from
Mangonui to about Tolaga Bay.   PROTEACEAE

544–545 Puataua, *Clematis forsteri*, is a small vine
found in lowland forest, along forest margins and
in lowland scrub throughout the North Island.
Sweet-scented flowers (fig. 545), 2.5–4 cm across,
occur from September to November, followed by
clusters of downy seeds from November to February
(fig. 544).   RANUNCULACEAE

543 Toru berries arise from December to March,
but persist throughout April and May, Taupo
(April)

544 Puataua seeds, Hinakura (January)

545 Puataua leaves and flowers, Hinakura (October)

548 Toro spray showing leaves and flowers clustered round the branch, Mt Holdsworth (November)

550 Toro branch with berries, Mt Holdsworth (December)

549 Close-up of flowers of toro, Mt Holdsworth (November)

546–547 Bush clematis/puawhanganga, *Clematis paniculata*, is a strong, woody liane that climbs to the tree tops to produce, in springtime, great clusters of white flowers (fig. 546), each flower softly fragrant and up to 10 cm across. Found throughout New Zealand in forest and along forest margins and in scrub. RANUNCULACEAE

548–550 Toro, *Myrsine salicina*, is a small canopy tree to 8 m high, with long, smooth, narrow, leathery leaves, 7–18 cm long by 2–3 cm wide. Flowers occur from August to January as dense-flowered fascicles along the stems (fig. 548), with each flower (fig. 549) about 3 mm across, followed from September to May by clusters of berry-like fruits (fig. 550). Found in lowland and montane forests from North Cape to about Greymouth. MYRSINACEAE

546 Bush clematis in flower, Renata Track, Tararua Range (November)

547 Close-up of flowers of bush clematis, Renata Track, Tararua Range (November)

The genus *Pittosporum* in New Zealand

Altogether there are 26 species of *Pittosporum* known from New Zealand; all are shrubs or small trees found in lowland and montane forests and lowland scrub, either throughout the country or confined to restricted localities as with *P. dallii*, found only in north-west Nelson. Some species have flowers borne singly, others are in panicles or umbels, and they may be monoecious or dioecious, and many are strongly sweet scented, especially in the evenings. Their leaves can vary from 5 mm to 15 cm in length, and from pinnatifid to elliptic in shape; some are aromatic. Most species flower between September and December, with seeds following from December onwards. A selection of these plants is shown here; others are already figured on pages 28, 29, 64, 67 and 122.   PITTOSPORACEAE

552   Tarata, female flowers, close up, Otari (October)

551–553 Tarata/lemonwood, *Pittosporum eugenioides*, is found throughout both North and South Islands. Berries (fig. 553) go black when mature.

551   Tarata, umbels of male flowers, Wilton's Bush, Wellington (October)

554–555 Black mapou, *Pittosporum tenuifolium* subsp. *colensoi*, is found from Opotiki southwards, and grows to 10 m high with leaves 4–10 cm by 2–5 cm wide.

554   Black mapou flowers, close-up, Erua (November)

556 Tawhirikaro/perching kohuhu, *Pittosporum cornifolium*, is found throughout both islands on rocks and also growing as an epiphyte on rata and other trees. It grows to 2 m high with young hairy branchlets and thin leathery leaves up to 7 cm long and 3 cm wide.

557 Golden-leaved kohuhu, *Pittosporum ellipticum*, is found from Mangonui to the Coromandel Ranges, and has golden tomentum covering young leaves, leaf undersides, petioles and peduncles.

553   Tarata berries, Wilton's Bush, Wellington (March)

555　Black mapou, flowers and leaves, Kaimanawa Mountains (September)

557　Golden-leaved kohuhu, flowers and leaves clustered round branch tip, Piha (September). Note the golden tomentum on the young leaves.

558　Pittosporum anomalum is a dwarf divaricating shrub found around Waitomo, the Volcanic Plateau and Arthur's Pass. Leaves are pinnatifid, 5–10 mm long and aromatic. The flowers, 6–8 mm across, are mostly solitary and terminal on branchlets.

559　Haekaro, *Pittosporum umbellatum*, is found from North Cape to Gisborne, as a tree to 8 m high. Flowers occur in many-flowered umbels from September to January. The leathery leaves are 5–10 cm long by 2 cm wide.

558　*Pittosporum anomalum* branch with flowers and leaves, Otari (September)

556　Tawhirikaro, flowers and leaves, Wilton's Bush (September)

559　Haekaro flowers, Ponatahi (September)

560 Ralph's kohuhu, *Pittosporum ralphii,* is found along streamsides and forest margins from Thames to Wanganui. Leaves are 7–12 cm long, densely clothed below with buff-coloured tomentum.

561 Dall's pittosporum, *Pittosporum dallii,* is found only near Boulder Lake on the Cobb Ridge and in isolated spots in north-west Nelson. It has coarsely serrate leaves up to 10 cm long by 3 cm wide. Sweet-scented flowers occur from October to January.

561 Dall's pittosporum flowers, Cobb Ridge (November)

560 Ralph's kohuhu flowers, Hukutaia Domain (November)

562–563 Monoao, *Halocarpus kirkii,* is a handsome canopy tree, 25 m high, not unlike a kauri when seen in the distance, and easily recognised by its juvenile foliage. Found occasionally in lowland forest to montane forests from Hokianga Harbour to the southern Coromandel Ranges and on Great Barrier Island. The female cone (fig. 562) is sometimes twinned and matures by April; the male cone (fig. 563) is mature in December.

PODOCARPACEAE

562 Monoao, female cone, Mt Kaitarakihi (April)

563 Male cone of monoao close up, Otari (December)

564 Kawakawa female flowers, close up, Waikanae (November)

565 Ripe fruit of kawakawa, Karaka Bay (February)

566 Kawakawa, male spike shedding pollen, Karaka Bay (September)

564–566 Kawakawa/pepper tree, *Macropiper excelsum*, is an aromatic shrub or small tree found in the undergrowth of lowland forests from North Cape to Westland and on the Chatham Islands. The zigzag branches, swollen at the nodes, bear broad leaves, 5–10 cm long by 6–12 cm across; the flowers and fruits are on spikes, 2.5–7.5 cm long, that arise all through the year.   PIPERACEAE

567 Hangehange flowers, close up, Wilton's Bush (October)

568 Green berries of hangehange, Akatarawa Saddle (February)

567–568 Hangehange/whangewhange/Maori privet, *Geniostoma ligustrifolium*, forms a bushy shrub or small tree, with brittle branches and shining pointed leaves, 7–9 cm long by 3–4 cm wide on petioles 10 mm long. Flowers (fig. 567), each about 5 mm across, arise in axillary cymes from September to November, and the berries, produced in profusion, persist from November to March or even as late as June (fig. 568).   LOGANIACEAE

569–571 Karapapa, *Alseuosmia pusilla* (figs 570–571) and **Northern karapapa,** *Alseuosmia banksii* (fig. 569), are both shrubs up to 2 m high found in undergrowth of lowland and montane forests from North Cape to Marlborough, generally in dense shaded areas. The leaves are of variable shape, 6–20 cm long, crenulated or toothed. The very strongly sweet-scented, trumpet-shaped flowers (fig. 570), 4 cm long, arise on drooping stalks beneath the leaves from April to November, and the berries (up to 9 mm long) ripen during February and March. ALSEUOSMIACEAE

569 Northern karapapa flowers and leaves, near Waipoua (April)

570 Karapapa flowers and leaves, Mt Holdsworth (June)

571 Karapapa berries, Mt Holdsworth (February)

572 Flowers of rata vine, New Plymouth (May)

572 Rata vine, *Metrosideros fulgens*, is a stout-stemmed liane found in lowland forests from the Three Kings Islands south to Westland. The thick leaves, 3.5–6 cm long by 10–25 mm wide, are on stout petioles, and the flowers occur in terminal cymes from February to June. MYRTACEAE

573 Large-leaved white rata, *Metrosideros albiflora*, is one of several white-flowering rata vines, with flowers occurring from November to March. *M. albiflora* has smooth, leathery leaves, 3.5–9 cm long by 2–3.5 cm wide, and is found in lowland forests, mostly kauri forests, in Northland. MYRTACEAE

574 Hairy-leaved white rata, showing leaves and flowers, Kaitoke Gorge (November)

574 Hairy-leaved white rata, *Metrosideros colensoi*, is a slender stemmed liane with leaves 15–20 mm long by 7–10 mm wide, the young leaves very hairy. It is found in lowland and coastal forests from Mangonui to Marlborough. Flowers arise as terminal and lateral small cymes from November to January.   MYRTACEAE

575 Small-leaved white rata, French Pass (January)

575 Small-leaved white rata, *Metrosideros perforata*, is usually a slender liane, but if exposed without support, this plant becomes a bushy shrub with entangled branches. Found in lowland forests, coastal forests and along forest margins from the Three Kings Islands to Banks Peninsula. The close-set, leathery leaves are 6–12 mm long by 5–9 mm wide. Flowers occur in profusion from January to March; they are normally white but pinkish and yellowish forms do occur.   MYRTACEAE

576 Small white rata spray with leaves and flowers, Lake Pounui (November)

573 Large-leaved white rata flowers, Mt Kaitarakihi (November)

576 Small white rata, *Metrosideros diffusa*, is a slender liane found in lowland forests from Northland to Nelson and Marlborough. The subsessile leaves, 7–15 mm long by 3–8 mm wide, are hairy when young, and the white or pinkish flowers occur from October till January.   MYRTACEAE

577   Shrubby rata flower, Canaan Track, Abel
       Tasman National Park (October)

577   Shrubby rata, *Metrosideros parkinsonii,* is a
straggly shrub to 7 m high, with four-sided
branchlets bearing leathery leaves, 3.5–5 cm long by
15–20 mm wide. The large, rounded flowers are produced in abundance along the branches, below the
leaves, from October to January. Found in lowland
and subalpine forests of the North Island, Great
Barrier Island and in the mountains of the Nelson
region.   MYRTACEAE

578–579   Rata/northern rata, *Metrosideros
robusta,* forms a large tree, to 30 m high, with four-
sided branchlets bearing leathery leaves, 2.5–5 cm
long by 1.5–2 cm wide, and large flowers (fig. 579)
produced in profusion on the canopy (fig. 578) from
November to January. Rata is found in lowland and
subalpine forests from the Three Kings Islands to
Nelson.   MYRTACEAE

578   Northern rata in flower,
       Lake Waikaremoana
       (January)

580   Flower of yellow-flowered
       southern rata, Denniston
       (January)

579   Flowers, close up, of northern rata, Pahiatua
       (December)

580–581   Southern rata, *Metrosideros umbellata,*
is a tree to 15 m high, with rounded branches
bearing narrow, thick, silky leaves, 5–7.5 cm long.
Flowers (fig. 581) occur in profusion all over the
tree from November to January. Yellow-flowered
forms of this tree (fig. 580) occur sparingly at Otira
and in Westland.   MYRTACEAE

582-584 Broadleaf/kapuka/papauma, *Griselinia littoralis*, and **Puka,** *Griselinia lucida*, have similar leaves and flowers, but whereas kapuka forms a round-headed tree to 15m high, puka starts life as an epiphyte on another tree, sending roots down to the ground to finally establish it as a separate tree. Both are found in lowland forests to subalpine scrub throughout New Zealand. The thick, leathery, glossy leaves of kapuka are normally equal sided at their bases (fig. 582), but in puka they are very lopsided. The tiny flowers, each 2-4 mm across (fig. 584), occur in panicles from October to December. The fruits (fig. 583), each up to 10 mm long, are black when mature and remain on the tree from December to August.   CORNACEAE

583 Puka spray showing berries and leaf shape, Woodside Gorge (April)

582 Kapuka spray showing flowers and leaves, Canaan Track, Abel Tasman National Park (November)

584 Kapuka flowers, close up, Renata Track, Tararua Range (November)

581 Spray of southern rata with flowers and leaves, Tautuku Beach (December)

585–588 Kaikomako, *Pennantia corymbosa*, is a small tree to 10 m high, which passes through a shrubby juvenile stage and has thick, leathery, sinuate-margined leaves, 5–10 cm long by 1–4 cm wide, on 10 mm long petioles (fig. 588). Found in lowland forests throughout New Zealand. Strongly scented flowers (fig. 586, male; fig. 587, female) occur in panicles (fig. 585), 4–8 cm long, from November to February. The drupes (fig. 588) ripen from February to May.   ICACINACEAE

585   Kaikomako showing panicles of flowers at branch tips, Lake Pounui (December)

586   Close-up of male kaikomako flowers, Lake Pounui (December)

588   Kaikomako drupes at the branch tips, Woodside Gorge (March)

587   Close-up of female kaikomako flower, Lake Pounui (December)

589–593 Kowhai, *Sophora microphylla, Sophora tetraptera, Sophora prostrata,* are all trees varying from 2 m to 12 m in height. The kowhai flower is New Zealand's national flower blooming from July to October, starting earlier in the north, later in the south and at higher altitudes. *S. microphylla* (fig. 589) occurs naturally throughout New Zealand, *S. tetraptera* (fig. 590) in the North Island from East Cape to the Wairarapa, both in lowland open forest, along forest margins and river banks, and in open damp and rocky places; *S. prostrata* (fig. 593) is found only in the South Island, mainly in open rocky places. All have feathery leaves, *S. microphylla* with 20–40 pairs of leaflets, *S. tetraptera* with 10–20 pairs and *S. prostrata* with 4–5 pairs.   FABACEAE

589 Kowhai flowers, *Sophora microphylla*, Peel Forest (October)

590 Kowhai flowers, *Sophora tetraptera*, Taupo (October)

591 Kowhai flowers, *S. tetraptera*, massed along a branch, Ponatahi (September)

592 Seeds of kowhai, *S. microphylla*, Woodside Gorge (January)

593 Kowhai flowers, *Sophora prostrata*, Woodside Gorge (September)

594–596 Rohutu, *Lophomyrtus obcordata*, is a spreading shrub to 5 m high, with rounded leaves, 5–10 mm across. Found in lowland forests from Mangonui southwards through both islands. The flowers (figs 594 & 596), about 6 mm across, occur from December to February, and the berries (fig. 595) ripen from April through May.

MYRTACEAE

594 Rohutu flowers and leaves, Rotorua (December)

595 Rohutu berries, Rotorua (May)

596 Rohutu, close-up of flower and leaves, Barton's Bush (December)

597 Pate flowers and leaves at branch tip, Akatarawa Saddle (February)

597–599 Pate/patete/five finger, *Schefflera digitata*, is a shrub or small tree to 8 m high, bearing compound leaves on petioles up to 25 cm long, and each made up of 3–9 thin leaflets, up to 20 cm long, with serrated margins. Found in lowland forests throughout New Zealand, usually in dampish places. The flowers, each about 7 mm across (fig. 599), arise in large, drooping panicles (fig. 597) below the leaves from January to March, and the berries (fig. 598) are coloured by March or April of the following year.   ARALIACEAE

600 Scented clematis, *Clematis foetida*, is a liane with strongly lemon-scented flowers produced in profusion from September to November. The plant sprawls over shrubs along lowland forest margins throughout New Zealand. The leaves are three foliolate with the leaflets sinuate or wavy margined.

RANUNCULACEAE

598 Pate berries, Lake Pounui (April)

599 Close-up of pate flower, Akatarawa Saddle (February)

601 Hinau, *Elaeocarpus hookerianus*, is a branching canopy tree, 15 m high, with leathery leaves, 10–12 cm long by 10–30 mm wide, having a silky tomentum on their undersides. Flowers similar to pokaka occur in great profusion from October to February, and the drupes, 8–12 mm long, mature from December to March but persist on the tree till May. Hinau is found in lowland forests from North Cape south to about Tuatapere.

ELAEOCARPACEAE

601 Hinau flowers, close up, Otari (November)

602 Westland quintinia, *Quintinia acutifolia*, is a bushy tree, to 12 m high, with wavy-margined leaves, 6–16 cm long by 3–5 cm wide, on 2 cm long petioles. Flowers, each about 6 mm across, occur in racemes during October and November. Found in lowland forests on both Little and Great Barrier Islands, around Coromandel Ranges and National Park to Taranaki and from Collingwood to Hokitika. Seeds ripen during December and January.

ESCALLONIACEAE

600 Scented clematis flowers with leaves, Hinakura (October)

602 Westland quintinia showing leaves and flowers, Lake Kanieri (November)

603   Tawherowhero flowers, close
up, Kauaeranga Valley
(December)

603–605   Tawherowhero, *Quintinia serrata,* is a
small, open-branching tree to 9 m high with narrow,
often blotchy, coarsely serrate leaves, 6–12.5 cm long
by 10–25 mm wide, usually wavy margined and on
petioles 2 cm long. Found in lowland forests from
Mangonui south to about Taumarunui. Flowers (fig.
603), in racemes (fig. 605), occur from October to
December, with seeds (fig. 604) maturing from
January to March.   ESCALLONIACEAE

604   Tawherowhero, newly set
seeds, Kauaeranga Valley
(December)

605   Tawherowhero spray with leaves and flower
racemes, Kauaeranga Valley (December)

606   Northern forest hebe, *Hebe diosmifolia,* is a
much-branched shrub, 1–6 m high, found in lowland
forests and scrub in Northland. Leaves are 10–30 mm
long by 3–6 mm wide, and the flowers arise in large
corymbose heads, either white or lavender coloured,
each flower about 8 mm across, during September
and October.   SCROPHULARIACEAE

606   Northern forest hebe, flowers and leaves,
Mangamuka Gorge (October)

607   Pokaka flowers, close up, Hinakura (November)

608   Pokaka branches heavy with
       flowers, Hinakura (November)

609   Pokaka branch with drupes
       and leaves, Hinakura
       (February)

607-609 Pokaka, *Elaeocarpus dentatus*, is a branching, rounded canopy tree, up to 12 m high, with leathery, serrate-margined leaves, 3–11 cm long by 10–30 mm wide. Flowers (fig. 607), each about 10 mm across, in drooping 8–12-flowered racemes, are borne in great profusion (fig. 608) from October to January. The purple-coloured drupe (fig. 609), 8 mm long, is mature from November to March. Found in lowland forests from Mangonui south to Stewart Island.   ELAEOCARPACEAE

610-612 Mahoe/whitey-wood, *Melicytus ramiflorus*, is a small, branching tree, to 10 m high, with serrated leaves, 5–15 cm long and 3–5 cm wide. Flowers, 3–4 mm across, arise on 10 mm long pedicels as fascicles from the leaf axils or directly from the branchlets (figs 610 & 612). The violet berries, 4–5 mm long, form clusters along the branches (fig. 611) from November to March. Mahoe is found in lowland and montane forests throughout New Zealand and the Kermadec Islands.   VIOLACEAE

610   Mahoe flowers cluster thickly around a branch,
       Karaka Bay (December)

611   Mahoe berries on branches with leaves, Karaka
       Bay (February)

612   Mahoe flowers, close up, Karaka Bay
       (December)

613 Large-leaved whitey-wood with berries, Waipoua Forest (March)

614 Spray of large-leaved whitey-wood with flowers, Waitakere Ranges (March)

615 Mahoe-wao showing branch heavy with berries and leaves, Waipunga Gorge (April)

613–614 Large-leaved whitey-wood, *Melicytus macrophyllus,* is a small tree to 6 m tall, with large, leathery, serrated leaves, up to 20 cm long and 10 cm wide, on petioles 2 cm long. Flowers (fig. 614), 6–7 mm across, and berries (fig. 613) are more sparse than in mahoe and occur round the same periods. Found in lowland to montane forests from Mangonui to Opotiki.   VIOLACEAE

615–619 Mahoe-wao, *Melicytus lanceolatus,* is a slender shrub or a small tree, 5–6 m high, with narrow, serrate leaves, 5–16 cm long by 5–30 mm wide. Flowers (fig. 616) arise as in mahoe from June to November, each about 5 mm across, and vary in colour from a rich yellow to a dark blackish brown (figs 617–619). Berries (fig. 615) are similar to those of mahoe and ripen from July to February, and can persist to the next flowering time. Found in lowland and montane forests and along forest margins from Cape Brett southwards throughout New Zealand.
VIOLACEAE

616 Mahoe-wao, branch with dense flowers, Hauhungaroa Range (September)

617 Flowers of mahoe-wao, close up, showing colour variation, Mt Ruapehu (October)

618 Close-up of yellow flower of mahoe-wao, Otari (September)

619 Close-up of flowers of mahoe-wao showing stalks and colour variation, Mt Ruapehu (October)

622 Dwarf bush nettle in flower, Wilton's Bush, Wellington (January)

622 Dwarf bush nettle, *Urtica incisa*, is a small herb, about 45 cm high, found in lowland and montane forests, along forest margins and in shaded open places throughout New Zealand. Leaves, up to 5 cm long, are on petioles up to 7 cm long, and bear occasional stinging hairs. Flowers, shown here, occur from September to February. URTICACEAE

620 Dianella/turutu/blue-berry/ink-berry, *Dianella nigra*, forms a lily-like plant growing to about 1 m high found in moist and dry forest and on dry, forested or scrub covered hillsides throughout New Zealand. Leaves are 25–60 cm, but sometimes up to 100 cm, long, by 10–15 mm wide. The plant is noted for its magnificent blue berries that arise from December till March from small inconspicuous whitish flowers with yellow stamens.

PHORMIACEAE

620 Dianella plant with berries, Lake Pounui (February)

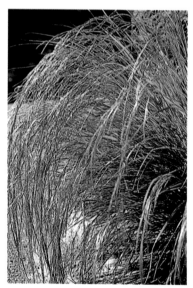

621 Bamboo grass in flower, Otari (January)

621 Bamboo grass, *Oryzopsis rigida*, is a graceful forest grass to 1 m high, found in lowland forests and along forest margins throughout New Zealand. Flowers occur in profusion during January.

GRAMINEAE

623–626 Turepo/milk tree, *Paratrophis microphylla*, forms a small tree, to 12 m high, found in lowland forests throughout New Zealand. Minute male flowers, 1–2 mm across, occur as spikes 10–20 mm long from October to February (fig. 624). Fig. 625 shows a minute female flower, about 2 mm across, which occurs on dense spikes up to 3 cm long from August to October. The flower-spikes of turepo and of towai are often affected by a fungal disease, which produces flower-like growths as shown in fig. 626. The leaves (fig. 623), 8–25 mm long by 5–12 mm wide, are crenate margined and on stout petioles 5 mm long. A white milky sap exudes from cut branches, leaves or twigs. From November to March the tree is adorned by brilliant red drupes (fig. 623).   MORACEAE

625   Female flowers of turepo, Waiorongomai (November)

626   Fungus disease affecting flowers of turepo, Wairongomai (November)

623   Turepo branch with drupes and leaves, Waiorongomai (February)

627 Towai/large-leaved milk tree, *Paratrophis banksii*, forms a spreading tree, to 12 m high, found in lowland forests from Mangonui south to the Marlborough Sounds. Except for the large crenate-margined, distinctly veined leaves, 3.5–8.5 cm long by 2–3.5 cm wide, on stout 10 mm long petioles, towai is very similar to turepo. Fig. 627 shows the brilliant drupes occuring from February to April.   MORACEAE

624   Male flowers of turepo, Waiorongomai (November)

627   Towai drupes, Waiorongomai (February)

628   Puka flower-head, Waikanae (May)

629   Puka flowers, close up, Waikanae (May)

630   Ripening puka fruits, Lake Pounui (March)

628–630   Puka, *Meryta sinclairii,* forms a handsome, round-headed tree, up to 8 m high, with large, glossy, thick, leathery leaves, 30–50 cm long by 15–25 cm wide on stout petioles 25–35 cm long. Flowers (fig. 628) arise as terminal panicles to 50 cm long from February to May. The succulent fruits are black when they mature 11–12 months later. Found on the Hen and Chickens and Three Kings Islands but now cultivated and grown in parks and gardens, sometimes also as street trees.   ARALIACEAE

631–632   Raukumara, *Brachyglottis perdicioides*, is a shrub to 2 m high found in lowland forests from East Cape to Mahia Peninsula. Thin coarsely serrated leaves are up to 5 cm long, and the strongly scented flowers (fig. 631), each to 10 mm across, occur from November to January. A hybrid formed by crossing with *Brachyglottis hectori* is grown in gardens and known as *Brachyglottis* 'Alfred Atkinson' (fig. 632).   ASTERACEAE

631   Raukumara in flower, Mahia (November)

632   *Brachyglottis* 'Alfred Atkinson' in flower, Otari (December)

633 Maire spray with drupes,
Hinakura (October)

634 Female flowers of maire, Hinakura (September)

635 Yellow drupes of maire, close up, Mt Ruapehu
(November)

633–636 Maire/black maire, *Nestegis cunning-hamii*, is a forest canopy tree to 20 m high, with narrow, leathery, rough-to-touch, willow-like leaves on stout petioles about 10 mm long. Minute flowers (fig. 636 shows a female) occur as racemes (fig. 634), 10–25 mm long, from September to December, and red or yellow drupes (figs 633 & 635) are mature from September to October of the following year. Maire is found in lowland forests from North Cape to Marlborough and Nelson.   OLEACEAE

636 Close-up of female flower of maire, Hinakura (September)

637 White maire drupes, Waipunga Gorge (March)

638 White maire flowers, Waipunga Gorge (December)

639-640 Bush pohuehue, *Muehlenbeckia australis,* is a scrambling vine that can completely cover trees, shrubs or rock faces in lowland forests throughout New Zealand. Leaves are 3–10 cm long, and the small flowers, each about 4 mm across, and the seeds (fig. 640) appear in panicles almost the year round.

POLYGONACEAE

639 Bush pohuehue in flower, Lake Pounui (January)

640 Bush pohuehue with seeds, Lake Pounui (April)

641 Mairehau flower-head and leaves, Rotorua (December)

642 Close-up of mairehau flower, Kauaeranga Valley (December)

637-638 White maire, *Nestegis lanceolata,* is a similar tree to black maire but smaller, to 15 m high, with smooth, leathery leaves, glossy above, 5–12 cm long by 10–35 mm wide. Minute flowers (fig. 638) occur as racemes about 2 cm long from November to January, and the drupes (fig. 637), which are more attenuated than those of black maire, mature about one year later.

OLEACEAE

641-642 Mairehau, *Phebalium nudum,* is an aromatic, branching shrub or small tree, to 3 m high, found in lowland forests from Mangonui to about Waihi. The narrow, finely crenulate, gland-dotted, leathery leaves, 2.5–4.5 cm long by 5–10 mm wide, are on short, twisted petioles. Flowers arise as corymbs (fig. 641), 5 cm across, with each flower 5–10 mm across (fig. 642), from October to December.

RUTACEAE

643–645 Tanekaha/celery pine, *Phyllocladus tri-chomanoides*, is a forest tree, to 20 m high, with the cladodes, which replace the leaves, arranged pinnately in two rows on rachides that arise in whorls along the branches. The cladodes, shaped like a celery leaf, are 10–25 mm long. Male catkins arise in bunches of 5–10 at the tips of the branches (fig. 645), and female cones arise arranged around the outsides of the cladodes (fig. 643), both from October to January. The seeds (fig. 644) take at least six months to mature. Found from North Cape to Nelson and Marlborough.   PODOCARPACEAE

643   Tanekaha, female cones at tip of branch, last season in outer ring, present season in inner group, Otari (October)

645   Mature male cones of tanekaha, Otari (October)

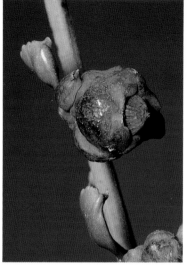

644   Mature female cone of tanekaha, Otari (April)

646   Kohurangi flowers and leaves, Mangamuka Gorge (January)

646 Kohurangi/kohuhurangi, *Urostemon kirkii*, grows either as an epiphyte or on the ground as a shrub or small tree, to 3 m high, in lowland and montane forests, more commonly from North Cape to the Coromandel Ranges but also sparingly throughout the North Island. The soft, fleshy, variable-shaped leaves, 4–10 cm long by 2–4 cm wide, are on petioles 10 mm long. Flowers, each about 5 cm across, occur in corymbs up to 30 cm across during December and January.

ASTERACEAE

647–650 Toatoa, *Phyllocladus glaucus*, is a tapering tree to 15 m high, with large, wedge-shaped, leathery cladodes (fig. 647), 4–6 cm long by 2–4 cm wide. Male cones (figs 648–649) arise in clusters on stout stalks at the branch tips (fig. 649), and the females (fig. 647) arise near the bases of the rachides during November and December. The seeds (fig. 650) mature some five to six months later. Found in lowland and montane forests from Mangonui south to a line from Taumarunui to Wairoa.

PODOCARPACEAE

647 Toatoa showing cladodes and ripe seeds, Waipahihi Botanical Reserve (May)

648 Toatoa, mature male cone with young leaf, Otari (November)

649 Toatoa spray with male cones, Otari (November)

651 Rautini with terminal panicle of flowers and rosettes of leaves, Otari (December)

650 Toatoa mature female cones showing protruding seeds, Taupo (May)

651 Rautini, *Senecio huntii,* is a shrub to 6 m high, with its branchlets marked by leaf scars. The long, narrow leaves, 5–10 cm long, occur mainly as rosettes of 20–30 leaves around the tips of the upturned branchlets. Flowers, each about 2 cm across, arise as dense terminal panicles, 12–18 cm high, from December to February. Found only on the Chatham Islands, in forests or on the edges of drier bogs, but now grown in parks and gardens.

ASTERACEAE

652 Black beech heavy with male flowers, Rimutaka Hill (October)

652 Black beech/tawhairauriki, *Nothofagus solandri*, forms a spreading tree, to 25 m high, with tough, leathery leaves, 10–15 mm long by 5–10 mm wide, clothed below with a dense whitish tomentum. In a good season male flowers are produced in profusion (fig. 652) from September to December. Black beech is found in lowland and montane forests from the Rotorua district south to Canterbury and Westland.   FAGACEAE

653 Mountain beech spray with male flowers, Volcanic Plateau (December)

654 Mountain beech seeds, Mt Ruapehu (February)

655 *Cyttaria gunnii* fungus on silver beech, Mt Robert Track (January)

655 *Cyttaria gunnii*, sometimes mistaken for a flower, is a fungus disease of beech trees, especially silver beech.

653–654 Mountain beech/tawhairauriki, *Nothofagus solandri* var. *cliffortioides*, is a smaller tree with more pointed leaves (fig. 654) than black beech, their margins rolled under and the lower surfaces pubescent. Male flowers (fig. 653) occur from November to January; female flowers are small and inconspicuous. Seeds (fig. 654) occur from February to April. Found in montane and subalpine forests and subalpine scrub from the Central Volcanic Plateau south but absent from Mt Taranaki and the Tararua Ranges.   FAGACEAE

656–658 Silver beech/tawhai, *Nothofagus menziesii*, is a tall tree, to 30 m high, with its branches arranged as if in tiers, and with round to oval, thick, leathery, crenate leaves (fig. 656), 6–15 mm long by 5–15 mm wide. Found in lowland and montane forests from Thames southwards, excluding Mt Taranaki. It can occur in subalpine scrub as a shrub or stunted small tree. Male flowers (fig. 656) and female flowers (fig. 657) occur from November to January, and the seeds, typical of beech trees, (fig. 658) mature from January to March.

FAGACEAE

656 Silver beech spray with male flowers, Tararua Forest Park (November)

657 Silver beech, female flowers, Eglinton Valley (December)

658 Silver beech seeds, Lewis Pass (January)

659–660 Hard beech/tawhairaunui, *Nothofagus truncata*, is a tall tree, to 30 m high, often with flanged buttresses to the trunk. Leaves (fig. 659) are 2.5–3.5 cm long by 2 cm wide, thick, leathery, smooth and coarsely serrated with a truncate apex. Flowers occur profusely (fig. 659) from September to December, and male flowers are often brilliant (fig. 660) as in *N. solandri*. Found in lowland and montane forests from Mangonui south to Kaikoura and Greymouth but absent from Mt Taranaki.

FAGACEAE

659 Hard beech spray, showing male flowers and leaves, Kaitoke (October)

660 Close-up of male flowers of hard beech, Woodside Gorge (January)

661 Red beech male flower, close up, Tararua Forest Park (November)

662 Male flowers and leaves of red beech, Hinakura (October)

661–663   Red beech/tawhairaunui, *Nothofagus fusca*, is a tall tree to 30 m high, often buttressed, with thin, leathery, coarsely and deeply serrate leaves, 2–4 cm long by 15–25 mm wide, on 4 mm long petioles (fig. 662). Flowers occur from September to December, male flowers (fig. 661), often brilliant red, carry huge quantities of pollen (fig. 663). Found in lowland and montane forests from Thames to Southland but absent from Mt Taranaki.

FAGACEAE

663 Red beech male flowers shedding pollen, Renata Ridge, Tararua Range (November)

664 Maire spray with leaves and racemes of flowers, Mt Meredith, Kaitaia (October)

665 Maire, pale coloured male flowers close up, Mt Meredith, Kaitaia (October)

666 Red-coloured female flowers of maire, Rewa Reserve, Kaitaia (October)

667 Fruits of maire, *M. salicifolia*, Rewa Reserve, Kaitaia (October)

664–667 Maire, *Mida salicifolia*, forms a slender tree, to 6 m high, which occurs locally in lowland forests from North Cape southwards but more commonly in the north. The leaves, 5–12 cm long by 3–10 mm wide, are entire, somewhat glossy, and usually arise alternately (fig. 664). Flowers, greenish or reddish tinged (fig. 665, male flowers; fig. 666, female flowers), occur from September to November, and the fruits mature from October to February.

SANTALACEAE

668–671 Kamahi, *Weinmannia racemosa*, is a spreading tree, to 25 m high, with coarsely serrate adult leaves, 3–10 cm long by 2–4 cm wide; juvenile trees have compound leaves with three leaflets and can bear flowers. Young leaves (fig. 670) are a rich red as they unfurl. Flowers on adult trees occur plentifully as racemes, to 12 cm long (fig. 668), from November to January, and the preceding reddish buds are very attractive (fig. 671). Found in lowland and montane forests from about Thames southwards.

CUNONIACEAE

670 Young leaves of kamahi, Tararua Forest Park (November)

668 Kamahi flowers, Akatarawa (November)

669 Close-up of kamahi flowers, Akatarawa (November)

671 Flower-buds of kamahi, Kauaeranga Valley (December)

672–674 Tawhero, *Weinmannia sylvicola*, is a small tree, to 15 m high, with persistent juvenile foliage of long, 5–10-foliolate leaves (fig. 673), and juvenile trees flower freely (fig. 673). Adult leaves are simple or 3–5 foliolate, leathery and serrate. Flowers (fig. 674) occur profusely as racemes, 8–12 cm long, from September to January. Tawhero is found in lowland forests from Mangonui south to the Rotorua district.   CUNONIACEAE

673 Adult form of tawhero showing terminal leaves and flowers, Kauaeranga Valley (January)

672 Juvenile form of tawhero showing terminal leaves and flowers, Kauaeranga Valley (January)

674 A small tree of tawhero in full flower, Mangamuka Gorge (January)

675 Kohia vine with fruits and leaves, Hinakura (May)

675–676 Kohia/New Zealand passion vine, *Tetrapathea tetrandra*, is a soft-wooded liane, reaching to 10 m high with alternate, entire, glossy, leathery leaves, 5–10 cm long by 2–3 cm wide, on petioles 2 cm long. Flowers (fig. 676) occur from October to December, and fruits (fig. 675) mature from March to May. Found in lowland forests, often along forest margins, from North Cape to Canterbury and Westland.   PASSIFLORACEAE

676 Kohia vine with flowers, Hinakura (November)

677–679 Bush lawyer/tataramoa, *Rubus cissoides*, is a liane, reaching to 15 m high, with interlacing, prickly branchlets and elliptic, serrate leaves in threes (fig. 677). Flowers arise as panicles up to 60 cm long (fig. 677), with male flowers (fig. 678) and female flowers (fig. 679) on separate lianes from September to November. Found in lowland and montane forests throughout New Zealand.   ROSACEAE

678 Bush lawyer, male flower and flower-buds, close up, Tararua Forest Park (November)

677 Bush lawyer in flower, Tararua Forest Park (November)

680 Tangled prickly branchlets of leafless lawyer, Woodside Gorge (January)

679 Bush lawyer, female flower, close up, Tararua Forest Park (November)

680–681 Leafless lawyer, *Rubus squarrosus*, is a liane that forms tangled masses of more or less leafless, prickly branchlets, the prickles recurved and yellow (fig. 680). It is found in lowland and montane forests and open, rocky places near forests throughout New Zealand. Flowers occur in panicles or racemes to 20 cm long from September to November, and the fruits (fig. 681), typical of lawyer fruits, mature from November to April. ROSACEAE

681 Fruits of leafless lawyer, Upper Ure River (March)

682   Puriri flowers and leaves, Wellington (April)

682–685   Puriri, *Vitex lucens*, is a massive spreading tree, to 20 m high, with four-angled branchlets and leaves made up of 3–5 glossy, undulate, leathery leaflets, each 5–12.5 cm long by 3–5 cm wide, with conspicuous veins, the whole on stout petioles to 10 cm long. Flowers arise in panicles of 4–15 each, and these and the drupes, about 2 cm in diameter, occur all the year round, with the most profuse flowering from June to October. Puriri grows naturally in lowland and coastal forests from North Cape to about Mahia and New Plymouth but is now extensively grown in parks and as a street tree.

VERBENACEAE

683   Puriri flowers, close up, Wellington (April)

684   Puriri berries, Wellington (February)

685   A puriri leaf, Wellington (March)

686-687 Tecomanthe, *Tecomanthe speciosa,* is a woody liane found naturally only on the Three Kings Islands but now grown in parks and gardens in frost-free districts. Flowers (fig. 687) occur from May to August, and fig. 686 shows the glossy, 3–5-foliolate leaves.   BIGNONIACEAE

686   Tecomanthe leaves, Hukutaia Domain (April)

688   Oro-oro male flowers after shedding pollen, Waipunga Gorge (October)

687   Tecomanthe flowers, Waikanae garden (June)

688-689   Oro-oro/narrow-leaved maire, *Nestegis montana,* is a small, much-branching, round-headed tree, 10–15 m high, found in lowland, montane and subalpine forests from Mangonui to Nelson and Marlborough. The very narrow-lanceolate leaves, 3.5–9 cm long by 6–9 mm wide, are leathery, slightly glossy, with a raised midrib. Slender racemes of flowers (fig. 688, male flowers) arise along the branches (fig. 689) from October to January, and the fruits ripen from December to March.
OLEACEAE

689   Oro-oro spray, showing narrow leaves and flower racemes, Waipunga Gorge (October)

PERCHING AND PARASITIC PLANTS

690–691 Kiekie/tawhara (edible bracts)/**ureure** (fruit), *Freycinetia banksii*, is a shrub that climbs on standing trees and fallen logs, with aerial roots that cling into fissures and crevices in the host bark. It sometimes grows on rocky ground or near water and is found from North Cape to Westland. The leaves, 1.5 m long by 2.5 cm wide, are spirally arranged, with the flowers (fig. 691) and fruits (fig. 690) hidden among the bases of the apical leaves. Flowers occur from September to November and fruits from January to February, ripening by May.   PANDANACEAE

691   Kiekie flowers, Lake Pounui (October)

690   Kiekie fruits set among whorls of leaves, Lake Pounui (April)

693   Kowharawhara female flower panicle, Mt Holdsworth (February)

692   Kowharawhara male flower panicle, Mt Holdsworth (February)

692–694 Kowharawhara/perching lily, *Astelia solandri*, is an epiphyte, mostly growing on branches and trunks of forest trees but occasionally on the ground. The drooping leaves, 1–2 m long by 2–3.5 cm wide, have three subequal nerves on either side of the midrib. Flowers in drooping panicles 15–40 cm long, males (fig. 692) yellowish white or maroon coloured, females (fig. 693–694) yellowish white to greenish, occur from October to June, with fruits present all the year round. ASPHODELACEAE

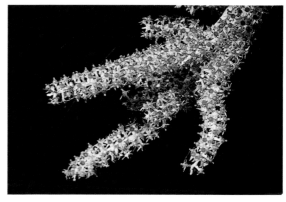

694   Close-up of female flowers of kowharawhara, Mt Holdsworth (February)

695-698 Kahakaha/perching lily, *Collospermum hastatum*, forms huge clumps on trees (fig. 695) or on rocks, and is generally similar to *A. solandri*. The leaves are 60 cm–1.7 m long by 3–7 cm wide, and the panicles of flowers, 15–30 cm long, appear from January to March, each extending out from a fan of leaves (fig. 696, female). Male flower panicles are similar, and male flowers are shown close up in fig. 698. The fruits (fig. 697) ripen from March till August, ultimately turning red. ASPHODELACEAE

695 Kahakaha plant on tree trunk, Waiorongomai (December)

696 Female flower panicle of kahakaha, Lake Pounui (November)

698 Close-up of male flowers of kahakaka, Lake Pounui (October)

697 Kahakaha fruits, Lake Pounui (October)

699 Wharawhara/shore astelia, *Astelia banksii*, is found on the forest floor of lowland and coastal forests of the North Island. It is similar to *A. solandri* in general appearance, with the leaves displaying several nerves on either side of the midrib and ascending for the lower half then drooping in the upper half. Flowers occur in panicles, up to 50 cm long, from March to June, with fruits (fig. 699), black when mature, being present the year round. ASPHODELACEAE

699 Wharawhara with panicle of ripe fruits, Lake Pounui (January)

700–701 Tree orchid, *Dendrobium cunninghamii*, is found throughout New Zealand on trees in lowland forests and scrub and occasionally on rocks. The long, thin, wiry, drooping stems bear narrow leaves, 3–5 cm long and 3 mm wide, and the strongly sweet-scented flowers, 2–2.5 cm across, arise in groups of 1–56 during December and January (fig. 701). The plant shown in fig. 700, growing on a cabbage tree in the south Wairarapa, was the largest *Dendrobium cunninghamii* I have ever seen, being almost 2 m high, but unfortunately it was destroyed during the 'Wahine storm' of 1965.

ORCHIDACEAE

700 Tree orchid in flower, Waiorongomai (December)

701 Tree orchid flowers, close up, Waiorongomai (December)

702 Hanging tree orchid/peka-a-waka, *Earina mucronata*, has drooping stems, 30 cm to 1 m long, bearing narrow, leathery, finely ribbed leaves up to 7.5 cm long. Sweet-scented flowers, about 6 mm across, occur as long drooping sprays during October and November. Found in lowland forests throughout New Zealand, preferably in shaded, somewhat damp places.

ORCHIDACEAE

702 Hanging tree orchid showing flowers and leaves, Hinakura (November)

703 Flowers of raupeka, Lake Pounui (February)

704–705 Red mistletoe/pirirangi, *Elytranthe tetra-petala*, is a bushy, branching, parasitic shrub, up to 1 m high and 2 m across (fig. 704), found growing on beech and *Quintinia* in lowland and montane forests from Dargaville southwards. The red flowers and buds (fig. 705) make a brilliant display from October to January.   LORANTHACEAE

706 Korukoru spray with flowers and leaves, Boulder Lake (December)

706 Korukoru, *Elytranthe colensoi*, is a smaller parasitic plant found on *Nothofagus* in the South Island, where it is more common, and on pohutukawa and *Pittosporum* in the North Island. The flowers occur from November to February and differ from those of *E. tetrapetala* by having their petals curled back on themselves.

LORANTHACEAE

704 Red mistletoe mass on a beech tree, Mt Ruapehu (January)

705 Red mistletoe flowers, close up, Mt Ruapehu (January)

703 Raupeka/Easter orchid, *Earina autumnalis* is a stout-stemmed, mostly erect, but occasionally drooping, orchid, found on trees and also on the ground in lowland forest throughout New Zealand. It has broader leaves than peka-a-waka and reaches 15–45 cm high. Sweet-scented flowers, 8 mm across, are produced at the tips of the stems from February to April.   ORCHIDACEAE

707 Common mistletoe spray with berries, Ure River Gorge (April)

707 Common mistletoe, *Loranthus micranthus*, is a bushy shrub attached to its host by a ball-like mass and found on many species of trees and shrubs in lowland forests and scrub throughout New Zealand. The small, greenish flowers are about 2.5 mm long and occur from October to December, being replaced from December to April by the conspicuous yellow berries shown in fig. 707.

LORANTHACEAE

PLANTS OF DRY RIVER BEDS

Many plants grow in these situations, forming mats and bushes that provide humus upon which, in time, other plants may grow.

708 Scabweeds The stone-strewn dry river and stream beds with their adjacent stony flats are normally covered by an assemblage of dense, mat-forming plants called scabweeds. These plants are often very beautiful and impart a splash of colour (fig. 708) to an otherwise drab, grey environment. They are also found in alpine dry, rocky places and screes. Except where indicated, they all belong to the family Asteraceae.

708   Scabweeds, Bealey River bed (January)

709   Mat daisy, *Raoulia parkii*, is a densely matted, prostrate, creeping and rooting daisy, found in South Island dry, rocky places. Flowers, 3–4 mm across, are produced in abundance during December and January.

709   Mat daisy in flower, Bealey River bed (January)

710   Golden scabweed, *Raoulia australis*, is found from the Tararua Range to Otago from sea-level to 1,000 m; it flowers in December and January.

711   Mossy scabweed with flowers, Bealey River bed (January)

711   Mossy scabweed, *Scleranthus uniflorus*, is a moss-like perennial herb found throughout the South Island in grasslands as well as in dry, rocky places. Tiny flowers occur from November to January.

CARYOPHYLLACEAE

713   Common scabweed, *Raoulia hookeri*, forms dense mats, up to 1 m across, varying in colour from green to grey, and is found in riverbeds from East Cape southwards. Flowers, 5–7 mm across, arise during December and January.

712   Tutahuna/green scabweed, *Raoulia tenuicaulis*, forms dense, silvery green to bright green mats, up to 1 m across, in gravel riverbeds and herbfields throughout New Zealand. The leaves are thick, 5 mm long by 2 mm wide, tapering to a pointed apex. White flowers, 6 mm across, arise from November to January, followed by silky seeds.

710   Golden scabweed in flower, Bealey River bed (January)

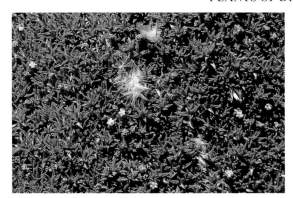

712 Tutahuna with flowers and seeds, Mt Taranaki (November)

713 Common scabweed in flower, Bealey River bed (January)

714 **Plateau scabweed,** *Raoulia hookeri* var. *albo-sericea,* is a greyish coloured scabweed, 60–70 cm across, flowering during December–January at the tips of erect branches with closely imbricate leaves. It grows on the Volcanic Plateau, Tararua Range and Ruahine Range.

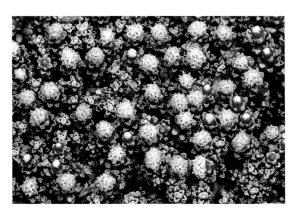

714 Plateau scabweed in flower, Mt Ruapehu (December)

715 **Green mat daisy,** *Raoulia haastii,* forms dense, green cushions, up to 1 m across and 30 cm high, and is found on the eastern side of the Southern Alps on riverbeds and river flats.

715 Green mat daisy, Bealey River bed (January)

716–717 **Wild Irishman/matagouri/tumatakuru,** *Discaria toumatou,* is a much-branched, stiff, spiny shrub (fig. 716), to 5 m high, found in dry riverbeds, open rocky places and sand dunes from coastal to subalpine regions throughout New Zealand. Flowers (fig. 717), 3–5 mm across, occur from October to January, followed by rounded berries from December to March. RHAMNACEAE

716 Matagouri spray showing small leaves, flowers and large thorns, Thomas River (December)

717 Close-up of matagouri flowers, Otari (October)

PLANTS OF DRY SCREES, ROCKY AND STONY PLACES

Many unusual and beautiful plants grow in dry, rocky or stony places in New Zealand, from about sea-level to high in the mountains. Often being exposed in precipitous situations, these are mostly very hardy plants.

718–719 Leafless clematis, *Clematis afoliata*, forms large masses of tangled stems, up to 1 m high, on open rocky ground from Hawke's Bay to Southland from sea-level to 1,000 m. Flowers, 1–2 cm across (fig. 718), in fascicles of 2–5 flowers each, occur from September to October, and the seed (fig. 719) sets during November and December.

RANUNCULACEAE

718   Leafless clematis in flower, Woodside Gorge (October)

719   Leafless clematis seeds, Otari (November)

721–722 Creeping lawyer, *Rubus parvus,* is a prostrate, creeping, rooting bramble with characteristic red bark and leaves 2.5–9 cm long. Flowers, 2–5 cm across, occur from September to November, and the juicy, edible fruits are ripe from November to April. Found in Nelson and Westland on stony ground and the stony banks of river valleys from about sea-level to 900 m.

ROSACEAE

720   Creeping pohuehue seed in capsule, Haast Pass (January)

720 Creeping pohuehue, *Muehlenbeckia axillaris*, is a prostrate, spreading, tangled shrub, forming patches, up to 1 m across, in subalpine rocky places, riverbeds and grasslands throughout New Zealand. Flowers, 3–4 mm across, occur singly and in pairs from November to April, filling the evening air with a rich, sweet scent. The black seeds, each about 3 mm long, sit in a white cup and are mature from December onwards.

POLYGONACEAE

721 Creeping lawyer, showing leaves and flower, Boulder Lake (November)

723 Porcupine plant with berries, Otari (February)

722 Creeping lawyer with fruits, Boulder Lake (February)

724 Close-up of flowers and leaves of porcupine plant, Otari (September)

723-725 Porcupine plant, *Melicytus alpinus,* is a stiff shrub, to 60 cm high and 1 m across, with rigid, interlacing branches bearing many lenticels and each terminating in a spine. Leaves are 6–18 mm long, and the tiny bell-shaped flowers occur in November, hanging in thousands along the branches. White berries, 5 mm across, ripen during February. This curious plant occurs east of the Southern Alps in exposed rocky places between 600 and 1,300 m.

VIOLACEAE

725 Porcupine plant, showing interlacing branches and flowers, Otari (October)

726 Leafless porcupine plant in flower, Otari (November)

726 Leafless porcupine plant, *Melicytus angustifolius*, is similar to *M. alpinus* but grows to 1.75 m high and is without leaves. The tiny flowers occur in thousands from August to October, and the berry, 3–4 mm long, with blue blotches on white, is ripe from September to January.   VIOLACEAE

727 Pink hebe flowers, Mt Terako (November)

727 Pink hebe, *Hebe raoulii*, forms a straggly shrub sprawling over dry rocks among the hills of Nelson, Marlborough and Canterbury from 200 to 900 m. Flowers occur abundantly from October through to March.   SCROPHULARIACEAE

729 Lake Tekapo willow herb, *Epilobium rostratum*, is a small, spreading herb with erect stems, 4–15 cm high, found along streambeds, on rocky outcrops and in grasslands from Arthur's Pass to Lake Wakatipu, being particularly common on the mountains around Lake Tekapo. Flowers, 3–5 mm across, occur from December to February.

The genus *Epilobium*

The genus *Epilobium* contains some 200 species worldwide, of which 50 are endemic to New Zealand. Many of the New Zealand species grow in dry, rocky places, including screes and riverbeds, and a few of our common species are shown here.

ONAGRACEAE

728 Glossy willow herb in flower, Cupola Basin (January)

728 Glossy willow herb, *Epilobium glabellum*, is abundant and common in mountain regions (100–1,800 m) from East Cape southwards, in riverbeds, stony subalpine places, grasslands and herbfields. Flowers, 5–6 mm across, occur from December to February.

729 Lake Tekapo willow herb in flower, Lake Tekapo (December)

730 Scree epilobium, *Epilobium pychnostachyum,* is a small, woody-stemmed herb with stems decumbent at the base, then erect, to 25 cm long. Leaves, 10–20 mm long by 2–4 mm wide, tinged with orange-red, are coarsely dentate. Flowers, 8–9 mm across, occur in the upper leaf axils from October to January. Found on rock screes in the Ruahine Range, North Island, and from Marlborough to Lake Wakatipu in the South Island.

733 Large-flowered mat daisy plant in flower, Homer Saddle (January)

730 The scree epilobium in flower, Fog Peak (January)

734 Close-up of flowers of large-flowered mat daisy, Renata Trig, Tararua Range (November)

731 Woody willow herb with flowers, Homer Cirque (January)

733–734 Large-flowered mat daisy, *Raoulia grandiflora,* is a spreading shrub forming cushions or mats (fig. 733), up to 15 cm across, on rocks in exposed places between 900 and 2,000 m from East Cape southwards. Flowers (fig. 734), 8–16 mm across, occur during December and January.

ASTERACEAE

731 Woody willow herb, *Epilobium novaezelandiae,* is a woody, much-branched herb, up to 25 cm high with denticulate leaves to 15 mm long, found from Mangonui southwards in subalpine riverbeds, grasslands and open stony ground. Flowers and fruits occur from November to February.

732 Red-leaved willow herb, *Epilobium crassum,* is a woody, creeping herb, rooting at the nodes; the thick, fleshy, leathery leaves, 3–4 cm long by 9–14 mm wide, are entire and basally dull green strongly tinged with red. Flowers and fruits occur from November to February, and seeds can persist on the plant until April. The plant is found from the Nelson mountains to Lake Wakatipu in subalpine rocky places, screes and herbfields.

732 Red-leaved willow herb with seeds, Cupola Basin (April)

735 Mountain sandalwood; the black and the red spines are leaves, the small, dark-coloured, grape-like clusters are male flowers, Jack's Pass (January)

736 Mountain sandalwood with fruits, Boulder Lake (March)

737 Mountain sandalwood, showing female flowers, Jack's Pass (January)

735-737 Mountain sandalwood, *Exocarpus bidwillii,* is a rigid, much-branched, spreading shrub, to 60 cm high, found in subalpine rocky places and open areas from the Nelson mountains south throughout the South Island. In fig. 735 the broad, triangular, black spines on the stems are leaves, the small grape-like clusters behind the leaves on the short lateral spikes off the main stems are the developing male flowers; female flowers are below the depressions in the spikes (fig. 737). Flowers occur during January and February; the nut-like fruits (fig. 736) appear from January to April.

SANTALACEAE

The genus *Helichrysum*

The genus *Helichrysum* contains several species with appressed, more or less triangular leaves, found in dry, rocky places among the mountains of the South Island; some of these are illustrated here.

ASTERACEAE

738-739 Yellow-flowered helichrysum, *Helichrysum microphyllum,* has flower-heads (fig. 738), 5 mm across, from November to March. It is a shrub (fig. 739), to 50 cm high, found in the mountains from Nelson to North Canterbury, between 500 and 1,300 m.

738 Spray of yellow-flowered helichrysum with flowers, Waipahihi Botanical Reserve (February)

739 Yellow-flowered helichrysum plant in flower, Mt Cupola (January)

740 Marlborough helichrysum, *Helichrysum coralloides,* is a shrub to 60 cm high, with leaves densely surrounded by soft, woolly hairs; flowers occur only occasionally in summer. Found on the Kaikoura Mountains.

740 Marlborough helichrysum, showing the very woolly stems, Otari (March)

741–742 Common helichrysum, *Helichrysum selago,* is a shrub to 30 cm high, flowering profusely during summer (fig. 742), the flower-heads (fig. 741) 6–7 mm across. Found in rocky places throughout the South Island mountains.

743 Hairy helichrysum, *Helichrysum plumeum,* forms a shrub, to 60 cm high completely covered by tangled, whitish hairs. Solitary terminal flowers occur during summer months. Found among the rocky mountains of West Canterbury and the Hunter Hills.

741 Common helichrysum flowers close up, Otari (November)

742 Common helichrysum plant in flower, Otari (November)

743 Hairy helichrysum showing hairy stems, Otari (March)

744   Everlasting daisy plant in flower, Dun Mountain
(November)

745   Close-up of flowers of everlasting daisy, Tauhara
Mountain (December)

744–745   Everlasting daisy, *Helichrysum belli-dioides,* is a prostrate, creeping and rooting shrub (fig. 744), with leaves, 5–6 mm long by 3–4 mm wide, clothed below with a soft, woolly tomentum. Flowers (fig. 745), 2–3 cm across, are produced on tall, woolly stems in great profusion from October to February. Found in lowland to subalpine rocky places, grasslands and open scrub throughout New Zealand.

Mountain hebes

Plants belonging to the genus *Hebe* are common in rocky places throughout the New Zealand mountains and a selection of these are illustrated here; all belong to the family SCROPHULARIACEAE.

746   Black-barked mountain hebe, *Hebe decumbens,* forms a decumbent shrub with shining, purplish black or dark brown bark and spreading, simple, slightly concave leaves, 12–20 mm long by 5–10 mm wide, with reddish-tinged margins. Flowers occur from November to February, and the plant is found among mountains from Nelson to Canterbury on rocky ledges at around 1,000–1,400 m.

746   Black-barked mountain hebe in flower, Mt Peel, Nelson (November)

747   Large-flowered hebe*, Hebe macrantha,* is a straggling, erect shrub, to 60 cm high, with thick, leathery leaves, 15–30 mm long, and flowers, about 18 mm across, occurring from December to February, Found between 800 and 1,600 m, in the mountains from Nelson to North Canterbury, on steep rock faces and in alpine scrub.

747   Large-flowered hebe flowers
with leaves, Wilberforce
River Gorge (January)

750   Thick-leaved hebe, *Hebe buchananii,* is a shrub, to 20 cm high, with spreading, slightly dished dull bluish-tinged, thick, leathery leaves, 3–7 mm long. Flowers occur from November to March, and the plant is found from the Godley Valley southwards, in the drier mountain areas.

748-749 Mt Arthur hebe, *Hebe albicans,* is a low, spreading shrub, 1 m high, with long, spreading, bluntly pointed, bluish green leaves (fig. 748), 15-30 mm long and 8-15 mm wide, found flowering from December to April (fig. 749) in rocky places among the mountains of Nelson. A similar species with blunter leaves, *H. amplexicaulis,* is found among the mountains of Canterbury.

751 Nelson mountain hebe plant with flower, Dun Mountain (December)

751 Nelson mountain hebe, *Hebe gibbsii,* is a sparingly branched shrub, to 30 cm high, with thick, red-edged, hairy-margined leaves, 10-18 mm long. Sessile flowers occur as terminal spikes from December to February. Found on the Nelson mountains, Ben Nevis and Mt Rintoul.

748 Mt Arthur hebe plant in flower, Mt Arthur (December)

749 Mt Arthur hebe flowers and leaves, close up, Mt Peel, Nelson (December)

752 Dish-leaved hebe spray, showing leaves and flowers, Cass (December)

750 Thick-leaved hebe spray with flowers, Lindis Pass (December)

752 Dish-leaved hebe, *Hebe treadwellii,* is a low, sprawling shrub with green, deeply concave, thick leaves 10-15 mm long and 5-10 mm wide, flowering during December and January and found in rocky places from the Victoria Range to Mt Cook.

753–754   Spiny whipcord, *Hebe ciliolata,* is a stiff shrub (fig. 754), to 30 cm high, with appressed leaves, found in rocky places among the Nelson and Canterbury mountains to 2,400 m. Terminal, sessile flowers (fig. 753) occur from November till January.

753   Spiny whipcord flowers, close up, Porter's Pass (November)

754   Spiny whipcord plant in flower, Porter's Pass (November)

756   Trailing whipcord, *Hebe haastii,* is a sprawling shrub with woody stems ascending at their tips, bearing imbricate, thick and leathery leaves, 6–13 mm long, with ciliated margins. Terminal flowers, each about 8 mm across, occur in terminal spikes from October to February, and the plant is found in rocky places to 2,200 m in the mountains from Nelson to Otago.

755   Spray of Colenso's hebe, showing flowers and leaves, Taruarau River (October)

756   Trailing whipcord stem, showing flowers and leaves, Mt Terako (September)

755   Colenso's hebe, *Hebe colensoi,* is a spreading, low-growing, North Island hebe, found in the headwaters of the Taruarau River on the Ruahine Range and in the Kaweka Range south to the Moawhango River. The leaves are blue, about 3 cm long and 1 cm wide. Flowers occur from August to November, both laterally and terminally on the branches.

757-758 Scree hebe, *Hebe epacridea*, is a low-growing, spreading shrub (fig. 757) with ascending stems bearing recurved leaves and terminal flowers (fig. 758). Flowers occur from November to February, and the plant is found on open rocky scree slopes among the mountains of Nelson, Marlborough and Canterbury.

759   Canterbury whipcord spray, *Hebe cheesemanii* showing leaves and flowers, Waterfall Valley, Cass River, Lake Tekapo (December)

757   The scree hebe plant in flower, Waterfall Valley, Cass River, Lake Tekapo (December)

759 Canterbury whipcords, *Hebe cheesemanii*, *Hebe tetrasticha*, are both shrubs, 20-30 cm high, with erect, leafy stems, 1-2.5 mm across, stouter in *tetrasticha*. Leaves in *tetrasticha* are longer than broad, shorter than broad in *cheesemanii*. Sessile flowers, in 1-3 pairs, in *tetrasticha* and 1-2 pairs in *cheesemanii* (fig. 759), occur from November to January. Both species are found in dry, rocky places in the mountains of Canterbury.

758   Stem of scree hebe showing leaves and flowers, Waterfall Valley, Cass River, Lake Tekapo (December)

760 New Zealand lilac, *Hebe hulkeana*, occurs naturally in dry, rocky places, to 900 m, over the northern half of the South Island; now extensively cultivated in parks and gardens. Branchlets are clothed with fine hairs, and the elliptic leaves, 7-10 cm long by 2-3 cm wide, have serrate or dentate margins. Flowers occur in profusion in crowded terminal spikes from October till December.

760   New Zealand lilac flowers, Otari (November)

761 Spray of blue-flowered
mountain hebe with flowers,
Upper Rangitata (December)

761 Blue-flowered mountain hebe, *Hebe pimeleoides*, is an erect, branching shrub, to 45 cm high, with narrow, lanceolate, bluish leaves, 5–15 mm long by 2–6 mm wide, found in the drier rocky places of the mountains from Marlborough southwards. Flowers, varying from bluish white to deep blue, occur in pairs on lateral hairy stalks from November to March.

Vegetable sheep

Vegetable sheep is the common name given to peculiar cushion-like daisy plants found in rocky places in subalpine and alpine regions of the New Zealand mountains, since these plants appear, in the distance, not unlike a flock of sheep. Two genera, *Haastia* and *Raoulia*, are involved.   ASTERACEAE

762 Tufted haastia, *Haastia sinclairii*, is a woolly, erect or decumbent shrub, sometimes forming mats. The leaves, 3.5 cm long by 15 mm wide, are covered by a thick, white, appressed tomentum; but plants in Fiordland can have a buff-coloured tomentum, and these are known as variety *fulvida*. Flowers, 2–3 cm across, occur from December to January. Found in subalpine and alpine screes, rocky places and fellfields on the eastern slopes of the mountains from Nelson to Fiordland.

762 Tufted haastia in flower,
Homer Saddle (January)

763 Red-flowered vegetable sheep, *Raoulia rubra*, is a daisy forming hard cushions up to 25 cm across and 15 cm high; found on the Tararua Range in the North Island and on the Paparoa, Haupiri and Mt Arthur Ranges in the South Island, on screes and in rocky places. Flowers, 2–3 mm across, occur during January and February.

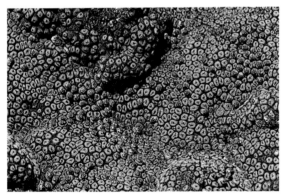

763 Red-flowered vegetable sheep with flowers, Mt
Holdsworth (February)

764 Silvery vegetable sheep, Mt Torlesse (November)

766 Giant vegetable sheep, close-up showing flowers, Cupola Basin, 1,710 m altitude (April)

767 Giant vegetable sheep; the seated figure gives a comparison of size, Cupola Basin, 1,710 m altitude (April)

764–765 Silvery vegetable sheep, *Raoulia mammillaris*, is a small plant, 50 cm across, with compacted silvery leaves found on rocks on Mts Torlesse (fig. 764), St Bernard, Hutt and Somers and the Craigieburn Range. Flowers, 2 mm across (fig. 765), occur from December to February, followed by downy seeds.

765 Silvery vegetable sheep, showing flowers, Fog Peak (December)

766–767 Giant vegetable sheep, *Haastia pulvinaris*, forms huge flat cushions, up to 2 m or more across (fig. 767, measured 8 square metres), and is found throughout the mountains of Nelson and Marlborough, at around 1,700 m altitude, on loose screes and rocky places, fellfields and the edges of more stable screes where the rocks are large. Flowers (fig. 766) occur during January and February, occasionally as late as April, followed by downy seeds.

768 Common vegetable sheep growing on rocks, Gunsight Pass, Cupola Basin, 1,900 m (December)

769 A large plant of common vegetable sheep, Cupola Basin, 1,700 m (April)

768–769 Common vegetable sheep, *Raoulia eximia*, forms tight, rounded cushions (fig. 769), up to 2 m across and 20–60 cm high. Found in dry, rocky places from 1,600 m to 2,700 m throughout the mountains of Nelson, Marlborough and Canterbury (fig. 768). The woolly leaves are tightly packed and overlapping, while the tiny flowers, 3 mm across, appear in January and February.

770   North Island edelweiss plant in flower, Mt Holdsworth (February)

770   North Island edelweiss, *Leucogenes leontopodium*, is a small, branching, woody plant with decumbent stems that have turned-up tips and sessile leaves, 8–20 mm long by 4–5 mm wide, entirely covered with silvery white wool. Flower-heads, 2.5 cm across, terminal on each stem, occur from November till March, and the plant, photographed on Mt Holdsworth, is found in subalpine and alpine rocky places, herbfields and fellfields.

ASTERACEAE

771   Black daisy plant in flower, scree on Fog Peak (January)

771   Black daisy, *Cotula atrata*, is a creeping, scree-inhabiting plant, found flowering during January and February between 1,250 and 2,000 m among the mountains of Marlborough and Canterbury. The flower-heads, 2 cm across, may be either black or brown.   ASTERACEAE

772   South Island edelweiss, *Leucogenes grandiceps,* is similar to *L. leontopodium* and is found in flower from November to February in rocky and stony places, up to 1,600 m, throughout the South Island mountains.   ASTERACEAE

773-774   Mountain cress, *Notothlaspi australe*, is a fleshy plant with grey-black, hairy leaves (fig. 773), 1–5 cm long, forming rosettes to 12 cm across, each rosette with many fragrant flowers (fig. 774), each 10 mm across, during December. Found on fairly stable alpine screes, between 1,600 and 1,800 m, among the mountains of Nelson and Marlborough.

CRUCIFERAE

773   Mountain cress plant on scree, Cupola Basin, 1,720 m altitude (April)

774   Mountain cress in flower, Cupola Basin, 1,600 m (December)

772   South Island edelweiss plant in flower, Homer Saddle (February)

775 Barbless bidibidi, *Acaena glabra*, is a creeping, prostrate, hairless plant, with branches, 50 cm long, rising at their tips and bearing leaves up to 5 cm long. Flower-heads, 2 cm across, arise from December to February. Found on screes, riverbeds and stony places throughout the eastern South Island.

ROSACEAE

776 Scarlet bidibidi plant with seed-heads, Volcanic Plateau, (January)

775 Barbless bidibidi plant with seed-heads, Mt Torlesse (December)

776–777 Scarlet bidibidi, *Acaena microphylla*, is a prostrate, spreading herb, forming bright red patches, to 75 cm across, when in flower from December to February. Found in stony stream beds, herbfields and grasslands of the Central Volcanic Plateau.

ROSACEAE

778 Penwiper plant, showing rosulate form, Sugarloaf, Cass (November)

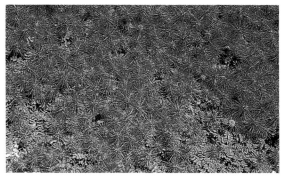

777 Scarlet bidibidi plant with two flowers and masses of seed-heads, Volcanic Plateau (January)

778–779 Penwiper plant, *Notothlaspi rosulatum*, is a fleshy herb, forming rosettes (fig. 778), 8–10 cm across and to 25 cm high when in flower. It is found on partially stabilised screes on the eastern side of the mountains from Marlborough to South Canterbury (fig. 779). Highly fragrant flowers occur during December and January.

CRUCIFERAE

779 Penwiper plant in flower and with seeds round the base of the flower-head, scree, Porter's Pass (January)

780   Fleshy lobelia, *Lobelia roughii*, is a spreading, hairless, fleshy herb found on screes and rock outcrops, between 900 and 1,800 m, in the mountains from Nelson to Otago. Flowers occur from October to February.   LOBELIACEAE

780   Fleshy lobelia in flower, scree on Fog Peak (January)

781   Grey-leaved succulent, *Lignocarpa carnosula*, forms a succulent plant, to 15 cm high, found on loose screes of Mt Torlesse and other mountains from Marlborough to Canterbury between 1,200 and 1,400 m. Tiny flowers, 2–3 mm across, occur from November to February.   APIACEAE

783   Snowy woollyhead, *Craspedia incana*, is a soft plant, covered all over with a silvery white wool, found on screes and rocks of the dry mountains of Canterbury. Leaves, 5–10 cm long by 2–3 cm wide, are arranged in rosettes, and the flower-heads, 2–3 cm wide, occur during January and February.   ASTERACEAE

781   Grey-leaved succulent in flower, scree on Fog Peak (January)

782   Mountain chickweed, *Stellaria roughii*, is a branching, succulent herb, with leaves 2 cm long by 3 mm wide, found on screes, between 1,000 and 2,000 m among the South Island mountains. Flowers, 2 cm across, having sepals longer than the petals, occur from December to February.   CARYOPHYLLACEAE

782   Mountain chickweed in flower, scree on Fog Peak (January)

783   Snowy woollyhead plant, showing dense wool on leaves and two opening flowers, scree on Porter's Pass (January)

784   Hairy pimelea, *Pimelea traversii*, forms a rather dense shrub, up to 60 cm high, found in rocky and stony montane to subalpine places throughout the South Island. Each individual flower of the flower-head has its perianth densely clothed with long silky hairs. The flowers are always pale pink and the thick leaves are faintly keeled.   THYMELAEACEAE

785 New Zealand chickweed, *Stellaria gracilenta,* is a stiff, erect herb, 10 cm high, with thick, awl-shaped leaves, 3–6 mm long, found in dry rocky places, fellfields and grasslands to 1,500 m from Mt Hikurangi to Fiordland. Flowers, 10–12 mm across, occur from October to March.

CARYOPHYLLACEAE

786   Rock cushion plant in flower among stones on Mt Lucretia (January)

785   New Zealand chickweed flowers, Cass (November)

787   Rock cushion, close-up to show flowers, Mt Robert (January)

786–787 Rock cushion, *Phyllachne colensoi,* forms hard cushions or mats to 40 cm across (fig. 786) in exposed rocky places in herbfields and fellfields to 1,850 m, from Mt Hikurangi south to Stewart Island. Tiny white flowers, with their anthers extending far beyond the lip of the corolla (fig. 787), occur during January and February.

STYLIDIACEAE

788 Arthur's Pass forget-me-not, *Myosotis explanata,* is a rosette-shaped herb, to 30 cm high, with leaves to 7 cm long, covered all over by soft, white hairs. Flowers occur as terminal cymes from December to February. Found in rocky places between 900 and 1,400 m, in the vicinity of Arthur's Pass.

BORAGINACEAE

784   Spray of hairy pimelea showing flower-heads, leaves and the very hairy basal parts of the flowers, Lake Lyndon (January)

788   Arthur's Pass forget-me-not plant in flower, Arthur's Pass (December)

789 Colenso's forget-me-not, *Myosotis colensoi*, is a creeping, prostrate perennial with lanceolate leaves, 2–3 cm long by 5–10 mm wide, more hairy on the upper surfaces. Flowers, 8 mm across, occur singly or in clusters from November through December, and the plant is found on limestone rocks around Broken River.   BORAGINACEAE

790–791 Small forget-me-not, *Myosotis monroi*, is a prostrate, rosette-forming plant with hairy leaves, found in rocky places on Dun Mountain, Nelson. Flowers occur as terminal cymes from late November to February.   BORAGINACEAE

789 Colenso's forget-me-not plant in flower, Castle Hill (November)

790 Small forget-me-not flowers, Dun Mountain (November)

791 Plant of small forget-me-not with flower-stalk, Dun Mountain (November)

Mountain ranunculi

Mountain ranunculi occur widely throughout the New Zealand mountains, with many species growing in rocky and stony places while others are found in grasslands, herbfields and fellfields. They are all conspicuous plants, and some of those found in rocky situations are depicted here.   RANUNCULACEAE

792 Haast's buttercup in flower, scree in Waterfall Valley, Lake Tekapo (December)

792 Haast's buttercup, *Ranunculus haastii*, has deeply divided leaves, to 15 cm long by 10 cm wide, and large flowers, 2–4 cm across, produced from November to January. *R. haastii* is found on screes in the mountains of Nelson and Canterbury.

793 Large feathery-leaved buttercup, *Ranunculus sericophyllus*, is found in the higher alpine rocky places and fellfields of the wetter regions from Lewis Pass to Northern Fiordland. Pictured here from above Wapiti Lake in Fiordland, the species is characterised by its deeply divided, hairy leaves on petioles 2–12 cm long, and the flowers, 2.5–4 cm across, produced from December to February.

794   Large white-flowered buttercup, *Ranunculus buchananii*, is found in rocky clefts between 1,500 and 2,300 m in the mountains from Lake Wakatipu to Lake Hauroko. The deeply divided leaves, to 15 cm long, are not hairy, and the flowers, 3–7 cm across, occur during December and January.

795–796   Korikori, *Ranunculus insignis*, is a branching hairy herb, to 90 cm high, found in rock crevices, herbfields and alpine grasslands from East Cape to the Kaikoura Mountains. Leaves, 10–16 cm wide, are thick and leathery. Flowers, 2–5 cm across, occur in profusion from November to February. Pictured in fig. 795 is the variety *glabratus* from Mt Ruapehu, while fig. 796 shows the parent form from Mt Holdsworth.

794   Large white-flowered buttercup in flower above Wapiti Lake, Fiordland, 1,600 m altitude (December)

795   Korikori plant in flower, Mt Ruapehu (January)

797   Snow buttercup, *Ranunculus nivicolus*, is an erect, hairy buttercup, producing flowering stems to 80 cm high, with flowers 3–5 cm across from September to December. Found on scoria slopes, in scrub and herbfields between 1,200 and 1,850 m, on Mt Taranaki, the central volcanoes, and the Kaweka and Raukumara Ranges.

796   Korikori flowers close up, Mt Holdsworth (December)

793   Large feathery-leaved buttercup in flower above Wapiti Lake, Fiordland, at 1,650 m (December)

797   Snow buttercup plant in flower, Mt Taranaki (September)

798   Blue-leaved ranunculus, *Ranunculus crithmifolius,* is a bluish black-leaved herb, to 80 cm high, found on screes above the Wairau River Gorge at about 2,100 m. Narrow-petalled flowers, about 2 cm across, occur during December and January.

798   Blue-leaved buttercup plant in flower, scree on Mt Dobson (December)

799   Small variable buttercup flowers and leaves, Porter's Pass (November)

799   Small variable buttercup, *Ranunculus enysii,* is a rosette-forming, hairless buttercup, with leaves either divided or lobed, 2–10 cm long by 10–70 mm wide, on grooved petioles to 10 cm long. Flowers, 15–30 mm across, occur during October and November. Found in rocky clefts or rocky sheltered places in tussock and scrub along the eastern mountains from North Canterbury to southwest Otago.

800   Haast's carrot plant in flower, Arthur's Pass (January)

800   Haast's carrot, *Anisotome haastii,* is up to 60 cm high, with purple stemmed, 2–3-pinnate, carrot-like leaves, 15–25 cm long by 6–12 cm wide, on stout petioles 8–12 cm long. Flowers occur from October till February, and the plant is found throughout the South Island mountains in rocky places and fellfields, between 500 and 1,600 m, mainly in western wetter regions.   APIACEAE

801   Snowball Spaniard, *Aciphylla congesta,* is a soft-leaved *Aciphylla* that forms compact masses of rosettes alongside cracks and water trickles that cross rocky outcrops and basins, between 1,300 and 1,600 m, in the mountains of west Otago and Fiordland. The flowers occur as white rounded umbels, up to 12 cm across, during December and January and appear like balls of snow on the rocks.

APIACEAE

801   Snowball Spaniard plant in flower above Wapiti Lake, Fiordland (December)

Mountain daisies

Mountain daisies belong to the genus *Celmisia* and are among the most common plants in the New Zealand mountains. They grow in herb- and fell-fields, on grasslands and screes or other stony places, and a selection of those found in rock situations are illustrated here.   ASTERACEAE

802 Fiordland rock daisy, *Celmisia inaccessa*, forms large mats, up to 1 m across, with lush, vivid green leaves, 2–6 cm long by 10–20 mm wide. Flowers, 5 cm across, occur during December and January, and the plant grows on limestone outcrops in the Wapiti Lake to Barrier Peaks area between the Doon and Stillwater Rivers.

802   Fiordland rock daisy plant in flower, Wapiti Lake, Fiordland (December)

804   Trailing celmisia, *Celmisia ramulosa*, has stiff, erect branches bearing thick, overlapping leaves with recurved margins. Flowers, 2–2.5 cm across, occur on woolly stalks during November and December. The plant is found trailing over rocks at about 1,400 m in the mountains of Otago, Southland and Fiordland.

804   Trailing celmisia in flower above Wapiti Lake, Fiordland (December)

803   White daisy flowers and leaves, Mt Ruapehu (October)

803   White daisy, *Celmisia incana*, is found in exposed rocky places, herbfields, fellfields and grasslands from the Coromandel Ranges to Otago. The thick, leathery leaves and flower stalks are densely clothed with appressed white hairs. Flowers, 2–3.5 cm across, occur from October to January.

805   Dusky Sound daisy plant in flower, Stillwater Basin, Fiordland (December)

805   Dusky Sound daisy, *Celmisia holosericea*, is found in rocky coastal to subalpine places throughout Fiordland. The shining leaves, 15–30 cm long by 4–6 cm wide, have silvery hairs on the lower surfaces, and the flowers, 5–7 cm across, occur in December and January.

806 Hector's daisy in flower and sprawling over rocks above Wapiti Lake, Fiordland (December)

807 Close-up of flowers and leaf rosettes of Hector's daisy, Homer Saddle, Fiordland (December)

806–807 Hector's daisy, *Celmisia hectori,* forms patches, to 1 m across (fig. 806), in rocky places or over rocks in herb- and fellfields from Arthur's Pass to Fiordland. The branches are clothed with leafy remains, and the hairy leaves are in rosettes, each of which bears a single flower (fig. 807), 2–2.5 cm across, during December and January.

808 Mountain rock daisy, *Celmisia walkeri,* forms large mats over rocks and rock-clefts in alpine regions southwards from the Spencer Mountains. The woody, 2 m long branches bear sticky leaves, up to 5 cm long by 5 mm wide, and the flowers, 2–4 cm across, occur from October to January.

809 Crag-loving daisy, *Celmisia philocremna,* forms a cushion, to 70 cm across, of thick, overlapping, succulent-like leaves 1.5–2.5 cm long by 5 mm wide, in crevices and on rock ledges between 900 and 1,800 m, in the Eyre Mountains. Hairy flower-stalks bear flowers, to 3 cm across, in December and January.

809 Crag-loving daisy plant with flower, photographed in the garden of Mr J. Anderson, Albury (December)

808 Mountain rock daisy plant in flower, Lake Ohau (November)

810 Fiordland mountain daisy, *Celmisia verbasci-folia,* is a large, tufted daisy, with thick, smooth, shining, leathery leaves, 15–25 cm long by 2.5–5 cm wide, having distinct veins above and velvety hairs below. Flowers, 2–2.5 cm across, occur on hairy stalks, 30–40 cm high, from December to February. Found in rocky places and herbfields among the mountains of West Otago and Fiordland.

810 Fiordland mountain daisy plant with flower, Wapiti Lake (December)

811–812 Spiny-leaved daisy, *Celmisia brevifolia,* forms large, loose, sprawling clumps, the woody branches clothed with leaf remains. The thick leaves, 10–15 mm long by 6–9 mm wide, have their margins bearing widely separated, small spines. Flowers, 2–3 cm across, occur on short, sticky stalks from October to January. Found in rocky places and herbfields, between 1,400 and 1,840 m, among the mountains of Canterbury and Otago.

811 Spiny-leaved daisy in flower, Old Man Range (October)

812 Spiny-leaved daisy, close-up to show spines on leaves, Old Man Range (October)

813 Bristly carrot plant with flowers, Temple Basin (January)

813 Bristly carrot, *Anisotome pilifera,* grows to 60 cm high with pinnate leaves, 10–30 cm long by 5–10 cm wide, on stout petioles to 10 cm long. Flowers, as compound umbels, arise from October to March, and the plant grows in rocky places in the South Island mountains, at about 1,850 m, from north-west Nelson southwards.   APIACEAE

814   Broad-leaved carrot, *Gingidium montanum,* is a stout aromatic herb, to 50 cm high, leaves radical, each with 5–7 pairs of serrate, sessile leaflets on 40 cm long, purplish petioles; terminal leaflets are usually 3–5 lobed. Flowers in umbels, 10 cm across, occur from October to January. Found in rocky and gravelly places and grasslands in coastal to subalpine regions from the central volcanoes southwards.

APIACEAE

815   Thread-like carrot with flowers and seeds, Mt Dobson (December)

815–817   Thread-like carrot, *Gingidium filifolium,* is a slender herb, to 30 cm high, with grooved, slender stems branching near their tops (fig. 815). Flowers are in umbels, 5–10 cm across, with male flowers (fig. 816) and females (fig. 817) in the same umbel from November to February. Found east of the Alps on scree and debris slopes from Dun Mountain to the Ben Ohau Range.   APIACEAE

818   Prostrate dwarf broom, *Carmichaelia enysii,* forms patches to 10 cm wide and 5 cm high on stony river terraces and grasslands at around 1,400 m on the eastern slopes of the Southern Alps from Arthur's Pass to Fiordland. Flowers occur from November to January.   FABACEAE

814   Broad-leaved carrot plant with flowers, Te Anau (January)

817   Thread-like carrot, close-up to show female flowers, Dun Mountain (November)

818   Prostrate dwarf broom, plants with flowers, Kurow (November)

819   Kopoti plant with flower, Dun Mountain (November)

816 Thread-like carrot, close-up of male flowers, Dun Mountain (November)

821 Cushion plant with flowers, Fog Peak (January)

819–820 Kopoti, *Anisotome aromatica*, grows throughout New Zealand in alpine and subalpine rocky places, grasslands, herb- and fellfields. Flowers occur as umbels from October till February, and plants can be 50 cm high at lower altitudes but only 10 cm in high alpine regions. APIACEAE

821 Cushion plant, *Pygmea pulvinaris* and *P. ciliolata*, both form a soft, moss-like, dense cushion, up to 10 cm across and about 4 cm high, in subalpine and alpine rocky and shingly places from Nelson to Canterbury. The tiny leaves, 2.5–4 mm long by 1 mm wide, are sparingly clothed with long, bristle-like hairs. Flowers, 5–6 mm across, occur in December and January. SCROPHULARIACEAE

820 Kopoti plant in full flower, Mt Ruapehu (January)

822 Variable cushion plant *Pygmea ciliolata*, in flower, Stillwater, Fiordland (December)

822 Variable cushion plant, *Pygmea ciliolata*, is similar to *P. pulvinaris* but rather variable in size and rigidity, with the young leaves ciliated along their margins. Found in bare rocky places and fellfields from north-west Nelson to Fiordland. Fig. 822 shows the variety *fiordensis*, from the Stillwater, in flower in December. SCROPHULARIACEAE

823 Auckland Island forget-me-not, *Myosotis capitata*, grows on the Auckland Islands, between sea-level and 600 m, on rocky clefts and ledges. The oblong to spathulate, hairy leaves, 3–12 cm long, are in rosettes. Flowers occur on hairy stems from November to February. BORAGINACEAE

823 Auckland Island forget-me-not plant in flower, Auckland Islands (February)

824 Chatham Island geranium, *Geranium traversii*, forms prostrate stems to 60 cm long with upturned tips. Both leaves and stems are densely clothed with silvery hairs, and the flowers, 2–2.5 cm across, are produced in abundance from October to February.

GERANIACEAE

825   Poor Knights lily plant with flowers, from a plant grown at Seatoun, Wellington, by the late Dr W. R. B. Oliver (November)

824   Chatham Island geranium flower and leaves, Otari (October)

825 Poor Knights lily/raupo-taranga, *Xeronema callistemon,* grows on the Poor Knights and Hen Islands only, in sunny, rocky situations between sealevel and 300 m. Leaves are 60–100 cm long by 3–5 cm wide, and the flowers appear on stout stalks, 25–30 cm long, from October to December.

LILIACEAE

826   Dusky Sound pimelea spray with flowers, Stillwater, Fiordland (December)

827   Northern snowberry in full flower, Mt Ruapehu (January)

826 Dusky Sound pimelea, *Pimelea gnidia*, is a much-branched shrub, to 1 m high, with dark red-brown bark and shining, oblong-lanceolate leaves, 1.5–2 cm long by 4–6 mm wide, on short petioles. Flowers with densely hairy perianth occur during December and January. Found in subalpine, stony scrub or herbfields from Coromandel Ranges to Fiordland.   THYMELAEACEAE

827 Northern snowberry, *Gaultheria colensoi*, is a sprawling, bushy shrub, to 60 cm high, with thick, generally crenulate margined leaves, 8–12 mm long by 5–9 mm wide, on petioles 1 mm long. Flowers, 2–3 mm long, occur in profusion from November to January. The berries, 3 mm across, ripen during February and March. Found in rocky places and grasslands among the North Island mountains from Mt Hikurangi southwards.   ERICACEAE

ALPINE AND SUBALPINE SCRUB

As one moves upwards out of the forest towards the open slopes of the New Zealand mountains there is usually a distinct zone of shrubs and small trees with some tussocks and herbaceous plants that forms a transition from the forest to the subalpine tussocklands, herbfields and fellfields. This zone constitutes the subalpine and alpine scrub.

828 Springy coprosma, *Coprosma colensoi*, is a slender, springy shrub, to 2 m high, with narrow, leathery leaves, 15–20 mm long by 5–8 mm wide, on hairy petioles 2–3 mm long. Flowers arise from September till December, and the elongated drupes are ripe by the following May–June. Found in subalpine scrub and forest from the Coromandel Ranges south to Stewart Island.   RUBIACEAE

828   Springy coprosma, showing leaves and drupes, Mt Holdsworth (June)

830   Musk tree daisy leaves and flowers, Mt Huxley (December)

829 Sprawling coprosma, *Coprosma cheesemanii*, is a sprawling, prostrate shrub with divaricating, pubescent branches, often forming a tight compact bush. Leaves are 8–11 mm long by 1–2 mm wide, and the drupes, 6–7 mm across, ripen during February and March. Found in subalpine scrub, tussock and grasslands, and sometimes in forest, fellfields and alpine bogs from East Cape south to Stewart Island.   RUBIACEAE

829   Sprawling coprosma showing spray heavy with drupes, Volcanic Plateau (February)

830 Musk tree daisy, *Olearia moschata*, is a much-branched shrub, to 4 m high, producing a strong, musk-like scent, especially during hot weather. The close-set leaves, 8–15 mm long by 5 mm wide, are glabrous or nearly so above, but have an appressed white tomentum below. Flowers occur from November to March and the plant is found in subalpine scrub around 1,500 m altitude from the Lewis Pass southwards. Hybridisation with other species occurs, and the form illustrated may be a hybrid.

ASTERACEAE

831–832 Rough-leaved tree daisy, *Olearia lacunosa,* is a shrub or small tree, up to 5 m high, with its branchlets, leaf petioles, leaf undersides and flower-stalks all clothed with a thick, brown or rust-coloured wool. Leaves, 7.5–17 cm long by 8–25 mm wide, have prominent midribs and veins imparting an alveolate appearance to their undersides (fig. 832). Flowers (fig. 831), 10 mm across, in corymbs, occur from November to February. Found in subalpine scrub and forest from the Tararua Ranges south to the mountains of Nelson and Marlborough.

ASTERACEAE

833–834 Hakeke, *Olearia ilicifolia,* is a musk-scented shrub or tree, to 5 m high, found in subalpine scrub or forest from the Bay of Plenty southwards. The strongly dentate leaves, 5–10 cm long by 10–20 mm wide, on petioles up to 2 cm long, have an appressed yellow tomentum on their undersides and are narrower in North Island plants than in South Island plants. Flowers, each about 15 mm across (fig. 834), arise in large corymbs from November till January. Hakeke often grows luxuriantly along stream banks and tracks or around clearings, and quite large specimens grow in the upper Hollyford Valley.

ASTERACEAE

831   Rough-leaved tree daisy flowers, close up, Mt Arthur (January)

833   Hakeke showing leaves and corymbs of flowers, Ruahine Range (December)

832   Rough-leaved tree daisy leaves, Mt Arthur (January)

834   Hakeke, close-up of flower, West Taupo (December)

835-836 Pigmy pine, *Lepidothamnus laxifolius,* is the smallest pine in the world, often fruiting in the juvenile stage, and forms a prostrate, trailing, creeping plant, making dense mats in subalpine and alpine scrub, herbfields and fellfields from the central volcanoes southwards. Spreading juvenile leaves are up to 12 mm long while adult leaves are 1-2 mm long, appressed to the stems. Male strobili (fig. 835) appear during November and December, with fruits (fig. 836) following through March and April.

PODOCARPACEAE

836   Pigmy pine plant with ripe seeds, Arthur's Pass (April)

835   Spray of pigmy pine with male cones, Arthur's Pass (November)

837-838 Hard-leaved tree daisy, *Olearia nummularifolia* and variety *cymbifolia* are shrubs, to 3 m high, with sticky branchlets clothed with white or yellowish, star-shaped hairs during the young or immature stage. The leaves are thick and leathery, 5-10 mm long by 4-6 mm wide in *nummularifolia* (fig. 837), and 6-14 mm long with margins recurved to the midrib in *cymbifolia* (fig. 838), giving the appearance of a swollen or fat leaf. Flowers, 3-5 mm across, occur from November to April except for *cymbifolia,* in which flowers seldom occur after January. The species is found in subalpine and alpine scrub from the Volcanic Plateau southwards, but the variety *cymbifolia* is found only throughout the South Island.

ASTERACEAE

837   Hard-leaved tree daisy, *Olearia nummularifolia,* spray with leaves and flowers, Mt Ruapehu (February)

838   Hard-leaved tree daisy, var. *cymbifolia,* spray showing leaves and flowers, Mt Terako (December)

839   Ruapehu hebe spray with flowers, Mt Ruapehu
(December)

839   Ruapehu hebe, *Hebe venustula*, is an erect
hebe found from Mt Hikurangi to Mt Ruapehu
among alpine scrub. Flowers occur in profusion as
racemes, 3–4 cm long, from December to February.
SCROPHULARIACEAE

840   Canterbury hebe in full flower, Arthur's Pass
(December)

840   Canterbury hebe, *Hebe canterburiensis*, forms
an erect shrub, to 1 m high, with narrow leaves,
7–17 mm long. It becomes smothered with white
flowers during December and January. Found in
subalpine scrub or tussock grassland on the Tararua
Ranges and the mountains of Nelson, Marlborough
and Canterbury.   SCROPHULARIACEAE

841   Western mountain koromiko. *Hebe gracillima*,
is a rather loosely branched shrub, to 2 m high, with
finely pubescent branchlets and narrow, lanceolate,
spreading leaves, 4 cm long by 6 mm wide. Flower
spikes occur laterally, from January to April, and
are longer than the leaves. Found in damp areas of
subalpine scrub, mainly on the western slopes, from
the Nelson mountains south to Arthur's Pass.
SCROPHULARIACEAE

841   Western mountain koromiko
in flower, Cobb Valley
(January)

842–843   Tupare/leatherwood, *Olearia colensoi*,
forms a closely branched shrub, up to 3 m high, with
the branches clothed in a fawn-coloured wool.
Leaves are thick, leathery and serrated, 8–20 cm long
by 3–6 cm wide, with a white, appressed tomentum
below. Flowers, 2–3 cm across, occur from
December to January. Tupare is found in subalpine
scrub from Mt Hikurangi southwards to Stewart
Island.   ASTERACEAE

842   Tupare in flower, Wharite Peak (December)

844 Variable coprosma, *Coprosma pseudocuneata,* is a shrub to 3 m high of variable form and habit, being reduced to 5–6 cm high in exposed situations. Found in subalpine scrub, forest, grasslands, herbfields and bogs from Mt Hikurangi to Southland. The thick, leathery leaves are 15–20 mm long by 2–6 mm wide, and the drupes, 5–6 mm long, are produced in profusion from February to May and may persist on the plants in alpine situations until the following December. RUBIACEAE

845 Mountain lycopodium with fruiting spikes, Mt Holdsworth (January)

845 Mountain lycopodium, *Lycopodium fastigiatum,* is a creeping plant found all over New Zealand in subalpine scrub and herbfields; the creeping stems extend for up to 2 m with ascending branches to 30 cm high. The fruiting bodies are 2.5–5 cm long, occurring singly at the tips of the branches from December through to April. LYCOPODIACEAE

844 Variable coprosma sprays with drupes, Temple Basin (April)

846 Club-moss whipcord stems and flowers, Garvie Mountains (December)

846 Club-moss whipcord, *Hebe lycopodioides,* is a stiff, erect, branching shrub, to 1 m high, with four-angled branchlets clothed by thick, appressed, triangular leaves, 1.5–2 mm long, with ciliated margins and each terminating with an apical cusp. Flowers occur during December and January.
SCROPHULARIACEAE

843 Tupare flowers close up, Wharite Peak (December)

847 Shrub groundsel, spray with flowers and leaves, Mt Holdsworth (January)

847 Shrub groundsel, *Brachyglottis elaeagnifolia,* is a shrub or small tree, to 3 m high, readily recognised by its grooved branches and the brown to dark brown, appressed hairs that clothe the branchlets, leaf petioles, flower stalks and lower leaf surfaces. The thick, leathery leaves, 10–12.5 cm long by 7.5 cm wide, have conspicuous veins, and the flowers, each 8 mm across, occur in panicles to 15 cm long during January and February.

ASTERACEAE

848 Kaikoura shrub groundsel, *Brachyglottis monroi,* is a shrub, to 1 m high, with serrated leaves, 2–4 cm long by 15 mm wide, found in subalpine scrub among the Kaikoura Ranges and the mountains of Nelson. Flowers arise from December to March.   ASTERACEAE

849 South Island shrub groundsel, *Brachyglottis revoluta,* forms a compact shrub, to 3 m high, with ribbed leaves, 3–6 cm long by 2–3 cm wide, white tomentum below and grooved petioles 10–20 mm long. Flowers, 2 cm across, occur from January to March, terminally, on ascending stalks up to 10 cm long. Found in subalpine scrub and fellfields in western Otago and Fiordland.   ASTERACEAE

849 South Island shrub groundsel, showing flowers and leaves, Homer Cirque (January)

850 Nelson mountain groundsel, *Brachyglottis laxifolia,* forms a laxly branched shrub to 1 m high with grey-green leaves. It smothers itself with flowers, 2 cm across, from December to February. Leaves are up to 6 cm long and 2 cm wide, their undersides and the branchlets densely clothed with white hairs. Found among the mountains of Nelson and Marlborough in subalpine scrub.   ASTERACEAE

848 Kaikoura shrub groundsel, plant with flowers, Otari (December)

850 Nelson mountain groundsel flowers and leaves, Marlborough mountains (December)

851 Thick-leaved shrub groundsel, spray of flowers and leaves, Mt Holdsworth (February)

851 Thick-leaved shrub groundsel, *Brachyglottis bidwillii*, is a tightly branched shrub to 1 m high, with thick, leathery, elliptic leaves, 2–2.5 cm long and 10–15 mm wide, both leaves below and branchlets densely clothed with soft, white to fawn-coloured hairs. Flowers, 15 mm across, occur in panicles from December to March, and the plant is found in subalpine scrub and fellfields from Mt Hikurangi southwards. ASTERACEAE

852 Reticulate coprosma, *Coprosma serrulata*, is easily recognised by its markedly reticulate, thick, leathery leaves, 4–7 cm long by 2.5–4 cm wide, and the white bark that falls in flakes. Found in subalpine scrub, forest, grasslands and herbfields throughout the South Island. The drupes, 7–8 mm long, make a fine display in the autumn. RUBIACEAE

852 Spray of the reticulate coprosma showing leaves and drupes, Arthur's Pass (April)

853–854 Coromandel groundsel, *Brachyglottis myrianthos*, is a sparingly branched shrub, up to 4 m high, with membraneous, very coarsely serrated leaves (fig. 854), 7–18 cm long, found on the Coromandel Ranges in subalpine scrub on valley sides. Flowers (fig. 853), up to 10 mm across, occur in large panicles (fig. 854) from November to January. ASTERACEAE

854 Spray of Coromandel groundsel with leaf and flowers, Kauaeranga Valley, Billy Goat Track (December)

853 Coromandel groundsel flower, close up, Kauaeranga Valley, Billy Goat Track (December)

855   Mountain wineberry flowers,
Volcanic Plateau (November)

855–856   Mountain wineberry, *Aristotelia fruticosa*,
forms a much-branched, tangled, rather rigid shrub,
to 2 m high, with ovate to ovate-oblong leaves (fig.
856), 5–7 mm long by 4–5 mm wide. Flowers (fig.
855), varying in colour from white to red, occur
singly or in 3–6-flowered cymes from October to
December, and the berries (fig. 856), 3–4 mm across,
ripen from November to April. Found in subalpine
scrub and forest, and fellfields, throughout New
Zealand.   ELAEOCARPACEAE

856   Mountain wineberry spray with berries, showing
growth habit, Outerere Stream, Volcanic Plateau
(December)

857   Needle-leaved mountain coprosma, *Coprosma
rugosa*, found in lowland, montane and subalpine
grasslands, scrublands and forest margins from Mt
Hikurangi southwards, forms an erect rather rigid
shrub, to 3 m high, often with divaricating branches
clothed with reddish bark. The thick, pointed,
narrow leaves are 10–14 mm long by 1–1.5 mm wide.
Flowers appear in October and November, with the
translucent drupes, 6–8 mm long, following from
February to April and sometimes persisting on the
plant through the winter.   RUBIACEAE

857   Needle-leaved mountain coprosma with drupes,
Arthur's Pass (April)

858   Mountain five finger, *Pseudopanax colensoi*,
is a shrub, up to 5 m high, found in subalpine scrub
and forest throughout New Zealand. Strongly sweet-
scented male and female flowers occur from May
to October, and the rather flat, rounded, black seeds
occur in large bunches, ripening by about the following March. Leaves are 3–5 foliolate with the
thick, leathery, sessile leaflets serrate on the upper
halves.   ARALIACEAE

858   Mountain five finger with male flowers and
flower-buds, Mt Taranaki (May)

859 Red mountain heath in flower, Cleddau Canyon (January)

860 Pepper tree flower, Mt Holdsworth (October)

859 Red mountain heath, *Archeria traversii*, is a shrub, 2–5 m high, with many erect or spreading branches. The leathery leaves are 7–12 mm long by 1.5–3 mm wide, Flowers occur as terminal racemes, 10–25 mm long, from December to February. The plant is found in subalpine scrub and forest throughout the South and Stewart Islands but is local in occurrence.   EPACRIDACEAE

861 Spray of pepper tree with berries, Maungutukutuku (April)

860–861 Pepper tree, *Pseudowintera axillaris*, is a shrub or small tree, up to 8 m high, found in subalpine scrub and forest and in lowland forests throughout New Zealand. The leaves, 6–10 cm long by 3–6 cm wide, are shining green on both surfaces, highly aromatic and pungent to taste. Flowers (fig. 860), 10 mm across, occur as axillary fascicles from September to December, and the fruits (fig. 861) ripen to red from October till January, persisting on the plant through to June.   WINTERACEAE

862 Varnished koromiko, *Hebe vernicosa*, is a semi-erect to prostrate shrub found in subalpine scrub and fellfields throughout the north-west Nelson and Marlborough regions. The leaves, 15 mm long, have a varnished appearance and their petioles twist so that, though they arise opposite and alternate, they do not immediately appear to do so. Flower-heads, 5 cm long, arise laterally towards the branch tips during January and February. This hebe also grows, to 1 m high, in the beech forests of the Nelson Lakes region.   SCROPHULARIACEAE

862 Varnished koromiko spray with flower panicle, Mt Robert track (January)

863   Tararua hebe plant in flower, Mt Holdsworth (February)

863   Tararua hebe, *Hebe evenosa*, is a spreading, branching hebe, to 1 m high, found in the Tararua Ranges. The smooth, thick leaves are 15–20 mm long; the flowers occur as crowded lateral spikes around the branch tips during January and February.

SCROPHULARIACEAE

864   Cypress-like hebe branches with flowers, Marlborough mountains (December)

864   Cypress-like hebe, *Hebe cupressoides*, forms a slender, densely branched, round shrub, to 2 m high, found in subalpine scrub and on river flats east of the Southern Alps from Marlborough to Otago. Flowers occur during December and January.

SCROPHULARIACEAE

865   Ochreous whipcord, *Hebe ochracea*, is an erect whipcord *Hebe*, to 30 cm high, with glossy green branches, ochreous towards their tips. Flowers occur as terminal spikes during December and January. Found in subalpine scrub of west Nelson mountains.

SCROPHULARIACEAE

866   Dish-leaved hebe, *Hebe pinguifolia*, is one of a complex of species with dish-shaped leaves occurring in subalpine-alpine scrub on the eastern side of the drier mountains from Nelson to Canterbury. Crowded, sessile flowers occur during December and January. Other species belonging to the complex are *H. buchananii* and *H. treadwellii*, both found in similar places.   SCROPHULARIACEAE

866   Dish-leaved hebe, *Hebe pinguifolia*, showing leaves and flowers, Jack's Pass (January)

865   Ochreous whipcord branches with flowers, Mt Peel, Nelson (December)

867   North-west Nelson hebe, *Hebe coarctata*, is a spreading shrub, to 1 m high, with arching branches tending to be decumbent, found along the edges of beech forest in the Nelson Lakes area and on the Brunner Range. Flowers occur in December and January.   SCROPHULARIACEAE

867   Spray of north-west Nelson hebe with flowers, Mt Robert (January)

Snowberries

Low shrubs belonging to the genera *Gaultheria* and *Pernettya* are found throughout the New Zealand mountains from subalpine scrub to the higher alpine fellfields. All species bear flowers and edible fruits called snowberries in profusion during spring, summer and autumn.   ERICACEAE

868–869   Snowberry, *Gaultheria antipoda*, is 30 cm to 2 m high, with branchlets clothed with a fine, silvery down mingled with black or yellow bristles. Thick, leathery leaves, 5–15 mm long by 3–15 mm wide, have conspicuous veins and bluntly serrated margins. Flowers (fig. 869) occur from October to February, and the snowberries (fig. 868) ripen from December to April. Found all through New Zealand in subalpine scrub, fellfields and rocky places.

870   Tararua snowberry plant in full flower, Mt Holdsworth (December)

871   Tararua snowberry flowers close up, Mt Holdsworth (December)

870–871   Tararua snowberry, *Gaultheria subcorymbosa*, forms a branching shrub, to 1 m high, which smothers itself with flowers (fig. 870). It is found in the Ruahine and Tararua Ranges and the mountains of Nelson. Leaves, 15–20 mm long by 5–7 mm wide, are thick and finely serrate-crenulate. Flowers occur as terminal and subterminal racemes from November till March, followed by snowberries from January till May.

868   Snowberry plant with berries, Volcanic Plateau (February)

869   Close-up of snowberry flowers and leaves, Taupo (October)

872 Weeping matipo flowers close up, Upper Travers Valley (August)

873 Weeping matipo branches with leaves and berries, Upper Travers Valley (April)

872–873 Weeping matipo, *Myrsine divaricata*, is a shrub or small tree, to 3 m high, found in subalpine scrub and forests, preferably where the ground is moist, throughout New Zealand. The obovate leaves, 5–15 mm long by 5–10 mm wide, are emarginate (fig. 873); flowers, 2–3 mm across, occur along the branches (fig. 872) from June till November, and the berries, 4–5 mm across, ripen from August to April.
MYRSINACEAE

874–876 Mountain tutu, *Coriaria pteridioides*, forms a small shrub, 60 cm high, with square, slender, hairy branchlets and narrow, lanceolate leaves, 15–25 mm long by 2–4 mm wide, each with a prominent pair of lateral veins (fig. 874). Flowers (fig. 875) occur as racemes to 5 cm long from October to February, and the black berries (fig. 876) are ripe between November and April. Found in the North Island only in subalpine scrub on the Central Volcanic Plateau and on Mt Taranaki.
CORIARIACEAE

874 Spray of mountain tutu with flowers, Volcanic Plateau (December)

875 Flowers of mountain tutu, close up, Mt Dobson (December)

876 Ripe berries of mountain tutu, Volcanic Plateau (February)

877   Dense tutu spray with flowers, Arthur's Pass (December)

878   Ripe berries of dense tutu, Haast Pass (January)

877–878   Dense tutu, *Coriaria angustissima*, forms large patches, to 50 cm high, spreading by branching rhizomes. The erect stems bear narrow leaves, 7–10 mm long (fig. 877), and racemes of flowers (fig. 877), 3–5 cm long, from November to February. The shining, intensely black berries follow the flowers till May. Found in subalpine scrub and rocky places, mainly on the west side of the mountains.

CORIARIACEAE

879   Feathery tutu, *Coriaria plumosa*, is a shrub with narrow, feathery-looking leaves, 6–10 cm long by 1.5–3 mm wide, found from Mts Hikurangi and Taranaki southwards to Stewart Island in subalpine scrub and grasslands, and along lowland and subalpine stream banks. Flowers occur from October till February and are followed by crimson-black berries from November till March.   CORIARIACEAE

880–881   Mountain totara, *Podocarpus nivalis*, is a much-branching, prostrate, spreading shrub, to 1.3 m high, with leathery, closely set, thick-margined, spirally arranged, rigid leaves, 5–15 mm long by 2–4 mm wide. Found in subalpine scrub and along subalpine forest margins from the Coromandel Ranges southwards but growing in lowland forests in Westland. Male cones (fig. 880) occur during November and December; female seeds (fig. 881), produced on separate plants, ripen from March to May.   PODOCARPACEAE

880   Mountain totara spray, showing male cones, Arthur's Pass (December)

879   Feathery tutu with flowers, Mt Belle, Homer Tunnel (December)

881   Mountain totara spray with seeds on red arils, Mt Ruapehu (February)

882–885 Mountain toatoa, *Phyllocladus asplenii-folius* var. *alpinus*, is a shrub or a small tree, up to 9 m high, in which leaves are replaced by modified branchlets called cladodes. These are 2.5–6 cm long, thick, leathery in texture and with thickened margins. Flowers are produced from October till January, the male as catkins, 6–8 mm long, in clusters at the tips of the branches (fig. 882), and the female as clusters imbedded in swollen carpidia along the edges of cladodes (fig. 883). The seeds are black, nut-like and held in a white cupule (figs 884–885). Found in sub-alpine scrub and forest from the southern Coromandel Ranges southwards; also found in lowland forest in south Westland.   PODOCARPACEAE

886–888 Bog pine/mountain pine, *Halocarpis bid-willii*, is a spreading shrub or small tree, up to 3.5 m high and 6 m across, with 1–2 mm long, leathery, thick, appressed leaves. Found in subalpine scrub from the Central Volcanic Plateau southwards and in lowland forest in south Westland and Stewart Island. Male (fig. 887) and female (figs 886 & 888) cones occur from November till March.

PODOCARPACEAE

884   Mountain toatoa seeds, Mt Ruapehu (February)

882   Mountain toatoa leaves and male cones, Cupola Basin (December)

883   Mountain toatoa branches with female flowers, Mt Ruapehu (November)

885   Mountain toatoa spray heavy with seeds, Mt Ruapehu (February)

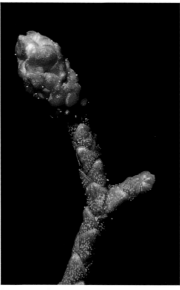

886   Bog pine spray with seeds, Lewis Pass (March)

887   Male cone of bog pine, Lewis Pass (November)

888   Female cone of bog pine with the seed on its white aril, Lewis Pass (March)

889   Mountain cottonwood, *Cassinia vauvilliersii,* is a shrub, to 3 m high, with grooved branches clothed with a brown-coloured, sticky wool. The leaves are leathery, 5–12 mm long by 1–2 mm wide, and clad below with a sticky, brown tomentum. Flowers occur as dense corymbs, 2.5–4 cm across, from December to April. Found in subalpine scrub and grasslands from about Thames southwards.

ASTERACEAE

890   Golden cottonwood, *Cassinia fulvida* var. *montana,* is a compact branching shrub, to 2 m high, with sticky branchlets clothed with a yellow-coloured tomentum, which imparts a golden colour to the shrub. The leaves, 4–8 mm long by 1 mm wide, are thick, leathery with revolute margins and each arising from an erect petiole. The flowers as in *C. vauvilliersii,* and the plant is found in similar situations from the Kaikoura Mountains southwards.

ASTERACEAE

889   Mountain cottonwood flower-head, Mt Taranaki (January)

890   Golden cottonwood spray with flower-head, Ure River bed (March)

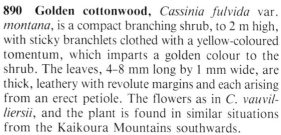
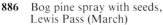

Subalpine grass trees

Grass trees belonging to the genus *Dracophyllum* are common in subalpine situations, some 18 of the 35 known New Zealand species occurring in mountain regions. A selection of these is illustrated on these two pages.   EPACRIDACEAE

891-892 Curved-leaf grass tree, *Dracophyllum recurvum*, is found on the Central Volcanic Plateau and the Ruahine and Kaimanawa Ranges among subalpine scrub and fellfields. Leaves are 10-40 mm long (fig. 891) and flowers (fig. 892) occur as spikes, 6 mm long, from December to February.

891   Curved-leaf grass tree on Mt Ruapehu (December)

892   Flowers of curved-leaf grass tree, Mt Ruapehu (December)

893 Turpentine scrub, *Dracophyllum uniflorum*, is an erect shrub, 1 m high, with pungent, needle-like leaves 12-25 mm long, with hairy upper surfaces and ciliate margins. Flowers occur singly around the tips of lateral branches during January and February. Found from the Kaimanawa Ranges southwards in subalpine scrub, herbfields, fellfields and grasslands.

893   Turpentine scrub in flower, Mt Holdsworth (February)

894 Spreading grass tree, *Dracophyllum menziesii*, is a spreading, branching shrub, often covering large areas in damp subalpine scrub or fellfields in the South Island. Leaves are 7-20 cm long by 10-20 mm wide, and flowers occur from December to February.

894   Spreading grass tree with flowers, Mt Belle, Homer Cirque (December)

895 Needle-leaf grass tree, *Dracophyllum fili-folium*, is a shrub or small tree, to 2 m high, with thread-like leaves, 6–16 cm long, with their tips three-sided. Flowers occur in December and January, and the shrub is found in subalpine and alpine scrub and fellfields of the North Island mountains.

897–898 Fiordland grass tree, *Dracophyllum fiordense*, is a small tree, to 3 m high, usually with only one trunk and a crown of leaves (fig. 897), each 60–70 cm long by 4–5 cm wide, beneath which are drooping panicles of flowers, 10–15 cm long (fig. 898), which occur during December and January. Found only in subalpine to alpine scrub in Fiordland and western Otago.

895 Needle-leaf grass tree leaves and flowers, Mount Holdsworth (December)

896 Trailing grass tree, *Dracophyllum pronum*, is a prostrate shrub with narrow, thick, leathery leaves, 5–12 mm long and 1 mm wide, swollen at the tips and with finely serrated margins. Flowers occur singly along the branchlets during December and January. Found from subalpine scrub to fellfields on the east side of the South Island mountains.

897 Fiordland grass trees in flower, Stillwater Basin, Fiordland (December)

896 Trailing grass tree spray with flowers, Waterfall Valley, Cass River, Lake Tekapo (December)

898 Flowers of Fiordland grass tree, Stillwater Basin, Fiordland (December)

899 Torlesse grass tree, *Dracophyllum acerosum,* is a slender, erect shrub, to 2 m high, found in subalpine scrub on the eastern slopes of the mountains between the Spencer Range and Arthur's Pass. Leaves are rigid, ascending, leathery, 2–3 cm long, and pungent. Flowers occur singly near the branch tips during January and February.

900   Specimens of the large grass tree, Mt Arthur (January)

899   Torlesse grass tree spray showing the thick leathery leaves and the flowers arising singly, Lewis Pass (January)

900–901 The large grass tree, *Dracophyllum traversii,* is the largest New Zealand grass tree, reaching a height of 10 m, with spreading, ascending branches bearing leaves 30–60 cm long by 4–5 cm wide (fig. 900). Flowers occur terminally on the branches as panicles, 10–20 cm long, during December and January, and the seeds ripen in a large, red-brown sheath (fig. 901) by late February. Found in subalpine scrub and forest from the Mt Arthur Tableland south to about Arthur's Pass.

901   A seed-head of the large grass tree, Otari (February)

TUSSOCKLANDS AND GRASSLANDS

Among the New Zealand mountains, tussocks and carpet grasses belonging to the family Gramineae clothe large areas of the rolling slopes above the bushline, forming meadows that can extend upwards to 2,000 m. In a healthy meadow the ground among the tussocks is carpeted with grasses, sedges, creeping mat plants, herbs, bluebells, bidibids, wild Spaniards, gentians and orchids. Some occasional subshrubs also occur here, and tussock meadows often merge above into herbfields or fellfields.

902   Narrow-leaved tussock in seed in Cupola Basin (April)

902   Narrow-leaved tussock, *Chionochloa pallens,* 60 cm–2 m high, is found abundantly from Mts Hikurangi and Taranaki southwards to altitudes of 1,700 m. It is probably the most common tussock among South Island mountains.   GRAMINEAE

903–904   Red tussock, *Chionochloa rubra,* to 1.6 m high, is found to 1,900 m altitude from East Cape and Mt Taranaki southwards (fig. 904) and is abundant in the South Island. Fig. 903 depicts a landscape of red tussock rolling across the Central Volcanic Plateau from Napoleon's Knob.

GRAMINEAE

903   Red tussock uplands above the Moawhango Gorge (April)

904   Red tussock plant, Lewis Pass (January)

905 Silver tussock, *Rytidosperma viride*, is up to 1 m high, with fine silvery leaves, and is found throughout New Zealand to 1,400 m altitude.

GRAMINEAE

905   Silver tussock plant with seeds, Tauhara Mountain (January)

908   Woollyhead plant in flower, Ward's Pass, Molesworth (January)

908 Woollyhead, *Craspedia uniflora*, is a soft-leaved perennial with leaves 5–12 cm long, their margins whitened by tangled, woolly hairs. Flower-stalks are woolly and bear flowers, 15–30 mm across, during December and January. Found from East Cape southwards in herbfields, fellfields and stony places.

ASTERACEAE

906 Carpet grass, *Chionochloa australis*, is found on steep slopes among the mountains of Nelson and Canterbury, up to 2,000 m, as tight, green mats. Flowers occur in December and January, followed by silvery seed-heads. Carpet grass is very slippery to walk on after snow or rain.

GRAMINEAE

906   Carpet grass in seed at Cupola Basin (April)

907 Blue bidibidi, *Acaena inermis*, forms a prostrate, spreading herb with bluish-coloured, 5–7-foliolate leaves up to 5 cm long. It is found in grasslands and fellfields and along the edges of screes among the South Island mountains up to 1,600 m. Flowers occur from December to January.

ROSACEAE

907   Blue bidibidi with flower-heads, Boulder Lake (February)

909 Red bidibidi/piripiri, *Acaena novae-zelandiae*, is a creeping and rooting, prostrate, silky-haired herb with creeping stems, up to 1 m long, forming large patches in subalpine grasslands. The globose flower-heads occur from September till February, and the seeds, which ripen from late November on, attach themselves to clothing and animal fur. Found throughout New Zealand in lowland and subalpine grasslands and open spaces.

ROSACEAE

910 Bidibidi, *Acaena viridior*, is similar to piripiri but never has red-tinged leaves. The flower-heads occur from September till February, and seed-heads from December on. The plant is found throughout New Zealand in lowland tussocklands or open spaces. ROSACEAE

910 Bidibidi plant with flower- and seed-heads, Hinakura (November)

911 Nelson mountain Spaniard, *Aciphylla ferox*, is similar in appearance to *A. horrida*, but smaller and found only among the mountains of Nelson and Marlborough, where it is common. APIACEAE

911 Plant of the Nelson mountain Spaniard in flower, Mt Travers (January)

909 Red bidibidi with seed-heads, Volcanic Plateau (November)

912 Bedstraw, *Galium perpusillum*, is a prostrate, spreading and rooting, perennial herb, forming patches to 30 cm across in damp situations between tussocks, in fellfields near streams or in partially shaded, open, rocky places. Three species of *Galium* occur in New Zealand, of which two grow in lowland areas; *G. perpusillum* is found in mountain regions. Small flowers, 2 mm across, occur in profusion from December to March. RUBIACEAE

912 Bedstraw plant in flower, Cupola Basin (December)

913 Onion-leaved orchid, *Microtis unifolia*, is 7–60 cm high, with fleshy stems and one leaf that is usually longer than the flower-spike. Flowers occur from October till February, either closely or sparingly spaced on the stem, and the plant occurs throughout New Zealand in grasslands and herbfields. ORCHIDACEAE

913 Onion-leaved orchid, close-up of spike showing detail of flowers, Volcanic Plateau (December)

914 Grassland orchid flowers, Volcanic Plateau (February)

915 Grassland orchid, red-flowered form, Ruahine Range (April)

916 Plants of odd-leaved orchid, Lewis Pass (December)

914–915 Grassland orchid, *Orthoceras strictum,* 20–60 cm high, is found in flower from December till April in herbfields and also in lowland dry, open grasslands from Northland to Nelson. Each flower-spike has 3–12 flowers, which may vary considerably in colour from spike to spike.

ORCHIDACEAE

916–917 Odd-leaved orchid, *Aporostylis bifolia,* has two leaves, with one leaf, 4–7.5 cm long, longer and broader than the other. The flower may be white or pink and is borne on a hairy stem from December to January. Found in subalpine grasslands from the Coromandel Ranges southwards to the Auckland and Campbell Islands, descending to sea-level in Stewart Island. ORCHIDACEAE

918 Rimu-roa flower, Lewis Pass (December)

917 Odd-leaved orchid flowers, close up, Lewis Pass (December)

918 Rimu-roa, *Wahlenbergia gracilis,* is a perennial herb to 40 cm high, with upright stems and very narrow leaves, 10–40 mm long, found in both sub-alpine and lowland grasslands throughout New Zealand. Flowers and fruits occur from September till April. CAMPANULACEAE

919-920 New Zealand bluebell, *Wahlenbergia albomarginata*, is a perennial, tufted herb with elliptic or spathulate leaves, 5-40 mm long by 1-10 mm wide, arranged as rosettes. Flowers, 10-30 mm across, may be erect, inclined or drooping and are borne on slender stems from November till February. Found throughout the South Island mountains in grasslands, herbfields and fellfields between 600 and 1,550 m altitude.

CAMPANULACEAE

920 New Zealand bluebell, close-up of plant and flowers, Tarndale, Molesworth (January)

919 New Zealand bluebell plant in flower, Mt Robert (January)

921 Marlborough bluebell, *Wahlenbergia trichogyna*, is an Australian plant that grows in New Zealand only in grasslands on the hills of Marlborough. Flowers occur in profusion from November to February. CAMPANULACEAE

922 Flowers of Maori bluebell, Volcanic Plateau (December)

921 Marlborough bluebell plant in flower, Otari (January)

922 Maori bluebell, *Wahlenbergia pygmaea*, forms a low-growing, hairless, tufted perennial herb with serrate-margined leaves, 15 mm long by 2-3 mm wide, arranged as rosettes. Flowers are held erect on slender stalks, 2-5 cm high, from November to February. Found in subalpine and alpine grasslands and herbfields from the Volcanic Plateau to Fiordland. CAMPANULACEAE

923–924 Benmore gentian, *Gentiana astonii*, is an erect, spreading plant, forming large plants up to 1 m across on limestone outcrops in montane to subalpine grasslands and herbfields along river valleys of the Kaikoura Coast between the Clarence and Ure Rivers. The leaves are in pairs, 2 cm long by 1–2 mm wide, and flowers, 2–3 cm across, occur singly in great profusion during March and April.

GENTIANACEAE

925 Grassland daisy, *Brachycome sinclairii* var. *sinclairii*, has spathulate, lobed leaves, 2.5–7.5 cm long, arranged in rosettes. Flowers 6–15 mm across on long scapes, 10–15 mm high, occur from October to December. Found in subalpine grasslands and herbfields from East Cape southwards.   ASTERACEAE

925   Grassland daisy plant in flower, Old Man Range (December)

923   Benmore gentian flowers, close up, Upper Ure River on slopes of Mt Benmore (March)

924   Plant of Benmore gentian in flower, slopes of Mt Benmore (March)

926   Prostrate snowberry flowers, close up, Mt Ruapehu (November)

926–927 Prostrate snowberry, *Pernettya macrostigma*, is a much-branching, prostrate, straggly shrub, to 20 cm high, forming patches in subalpine grasslands, herbfields, fellfields, and rocky places throughout New Zealand. Flowers (fig. 926), about 3.3 cm long, appear from November to January; fleshy berries, 4–7 mm across, each within a fleshy calyx matching the berry in colour (fig. 927), are ripe from December to May. *Pernettya* and *Gaultheria* species hybridise to produce *Gaultheria*-like plants with white or red berries.   ERICACEAE

927   Prostrate snowberry hybrid with red berries, Outerere Gorge, Volcanic Plateau (May)

Wild Spaniards or speargrasses

These are the popular names given to plants of the genus *Aciphylla*, found commonly from sea-level to 1,850 m throughout New Zealand. Thirty-nine species of *Aciphylla* are known between the coast and the high mountains, all characterised by their pinnate, sword-like or spear-like, rigid, pointed leaves. *Aciphylla* species are found in grasslands, herbfields, subalpine and alpine scrub and rocky places. A selection of the alpine species is shown here.   APIACEAE

928–929 Horrid Spaniard, *Aciphylla horrida,* forms clumps to 1 m high with flowers, to 1.5 m high (fig. 929), during December and January and is found in alpine scrub and herbfields on the eastern, dry side of the Main Divide in the South Island from Arthur's Pass to Fiordland. A similar but smaller speargrass, *A. ferox,* occurs among the mountains of Nelson and Marlborough.

928   Horrid Spaniard plants in flower, Mt Belle, Homer Tunnel (December)

929   A plant of the horrid Spaniard in flower, Mt Belle, Homer Tunnel (December)

930 Feathery Spaniard, *Aciphylla squarrosa* var. *flaccida,* is a soft-leaved form of the common Spaniard, *A. squarrosa,* flowering from December to February and found only on North Island mountains in damp alpine and subalpine scrub.

930   Feathery Spaniard plants in flower, Mt Holdsworth (February)

931   Subalpine Spaniard plants in flower, Wapiti Lake, Fiordland (December)

931 Subalpine Spaniard, *Aciphylla pinnatifida,* is a small Spaniard with leaves to 20 cm long, each conspicuously marked by a deep yellow-coloured, central stripe. Flowers occur during December and January. Found in alpine herbfields of western Otago and Fiordland.

932 Plants of the giant Spaniard in flower, Tasman Glacier morain (December)

933 Pigmy speargrass plant in flower, Waterfall Valley, Cass River, Lake Tekapo (December)

934 Close-up of flowers and leaf of pigmy speargrass, Waterfall Valley, Cass River, Lake Tekapo (December)

932 The **giant Spaniard**, *Aciphylla scott-thomsonii*, forms a large clump to 3 m high, bearing flower-heads on stalks, to 4 m tall, during December and January. Found in subalpine scrub and fellfields from about Mt Cook southwards on the drier side of the mountains.

933–934 **Pigmy speargrass**, *Aciphylla monroi*, is a small, tufted speargrass (fig. 933), found in subalpine grasslands, herbfields and rocky places up to 1,700 m among the mountains of Nelson, Marlborough and North Canterbury. Flowers (fig. 934) occur during December and January.

935 **Armstrong's speargrass**, *Aciphylla montana*, is a small speargrass found only among the mountains of the South Island from the Arrowsmith Range south to the Harris Mountains. Flowers on stalks, to 60 cm high, occur during December and January. A similar speargrass, *A. lyallii*, is found among the mountains of Fiordland.

936 Tararua speargrass plant in flower, Mt Holdsworth (December)

935 Armstrong's speargrass plant in flower, bank of Godley River, Lake Tekapo (December)

936 **Tararua speargrass**, *Aciphylla dissecta*, is a small plant, up to 40 cm high, found only in subalpine grasslands on the Tararua Range. Flowers occur during November and December.

937 Wild Spaniard, *Aciphylla colensoi,* is a large plant, to 1 m high, with distinct bluish-coloured, rigid, sword-like, pungent leaves, 30–50 cm long. Sweet-scented flowers occur on long pole-like stalks, 2.5 m high, arising from the centre of the plant, from November to February. Found in subalpine and alpine grasslands and herbfields from East Cape to Canterbury.

937 Wild Spaniard in flower, Mt Holdsworth (February)

938 Little mountain heath, *Pentachondra pumila,* is a slender dwarf shrub, forming patches to 40 cm across in subalpine and alpine grasslands, herbfields, fellfields and along the edges of alpine bogs, occasionally also in lowland grasslands and sand dunes. Flowers occur from November to February, and the large berries, 6–12 mm across, occur from December to April, often with the flower corollas remaining attached.   EPACRIDACEAE

938 Little mountain heath plant with flowers and berries, Jack's Pass (January)

939–941 Prostrate coprosma, *Coprosma petrei, Coprosma pumila,* are creeping and rooting coprosmas, forming large, dense, flat mats among grasslands, herbfields and stony places in subalpine regions throughout New Zealand. Flowers are erect and conspicuous during November and December (fig. 939, *C. petrei*). The drupes, 6–8 mm long, are translucent, ripe by March and persist on the plants until December (fig. 940, *C. petrei* & fig. 941, *C. pumila*); they may be greenish white, green, pale blue, orange-red, purplish red or claret in colour.
RUBIACEAE

939 Prostrate coprosma, *Coprosma petrei,* plant in flower, Key Summit (January)

940 Prostrate *C. petrei* plant with drupes, Molesworth (March)

941 Prostrate coprosma, *Coprosma pumila,* plant with drupes, Cupola Basin (December)

942–943 Brown-stemmed coprosma, *Coprosma acerosa* var. *brunnea*, is a sprawling plant with interlacing branches forming flattened mats, to 2 m across, in subalpine grasslands, especially along river terraces and among rocks between 600 and 2,000 m throughout New Zealand. Flowers occur from August to October and the translucent drupes, 5–6 mm long, ripen during March and April, those of the North Island being pale blue with darker stripes (fig. 942), those of the South Island a deep rich blue (fig. 943). RUBIACEAE

942 Brown-stemmed coprosma showing interlacing branches, leaves and pale drupes, Upper Waipunga River (April)

943 Brown-stemmed coprosma with deep blue drupes, Jack's Pass (March)

Maori onion, *Bulbinella* species

944–948 Maori onion is the popular name for six species of perennial lilies, 30–60 cm high, found in subalpine grasslands and herbfields from Lake Taupo and Mt Taranaki southwards. Three species are illustrated here; all have conspicuous yellow flowers, which occur from October to January, and in the South Island these plants often grow in great numbers over wide areas, making a brilliant display (fig. 944) when in bloom. Flower stalks can reach 45 cm high, with racemes to 15 cm long and flowers, each 10–14 mm across, occupying the upper third. *B. hookeri* is found from the Volcanic Plateau south to the Waiau River. *B angustifolia* occurs from south of the Waiau River to Southland. *B. gibbsii* var. *balanifera* is a shorter plant than *B. hookeri* and is found along the west side of the Southern Alps into Fiordland. ASPHODELACEAE

944 Maori onion, *Bulbinella hookeri*, clothes the Acheron River Valley floor (January)

945 Maori onion, *B. hookeri*, plant in flower, Acheron River Valley (January)

946 *B. hookeri* flowers, close up, Acheron River Valley (January)

948 Maori onion, *Bulbinella angustifolia*, flowers, Thomas River (December)

947 Maori onion, *Bulbinella gibbsii* var. *balanifera*, plant in flower, Wapiti Lake, Fiordland (December)

New Zealand subalpine orchids

Many species of orchids occur in our subalpine and alpine regions, but only two of the more striking species are illustrated here. ORCHIDACEAE

949 Broad-leaved thelymitra, *Thelymitra decora*, is up to 50 cm high, with a single, keeled, broad leaf, 10 mm wide; flowers occur in racemes of 1–10 flowers during November and December. Found throughout New Zealand in grasslands and herbfields to 1,250 m altitude. ORCHIDACEAE

949 Broad-leaved thelymitra flower, Volcanic Plateau (December)

950   Long-leaved thelymitra flowers, Volcanic Plateau (November)

951   Long-leaved thelymitra flowers, close up, Kauaeranga Valley, Billy Goat Track (December)

952   Tall mountain sedge in flower, Boulder Lake (January)

950–951   Long-leaved thelymitra, *Thelymitra longifolia*, (fig. 950), is 7–45 cm high with a single, fleshy, keeled leaf, 3–18 mm wide, found in grasslands and herbfields throughout New Zealand. Flowers, 8–18 mm across (fig. 951), occur during November and December and may be white, blue or pink in colour.   ORCHIDACEAE

952   Tall mountain sedge, *Gahnia rigida*, has erect leaves with long, drooping tips and inrolled margins. Flower-stems, 60 cm–2 m high, occur during January and February, and the plant is found in subalpine regions among the mountains of Nelson and Marlborough.   CYPERACEAE

953   Tall tufted sedge, plume of seeds, Mt Ruapehu (May)

953   Tall tufted sedge, *Gahnia procera*, is up to 1 m high, with flower-stems, 60 cm–1 m long, bearing panicles of flowers from October to December. Seeds, each 6 mm long (fig. 953), ripen from February to May. Found in subalpine and lowland grasslands from North Cape to Westland.
                                                CYPERACEAE

954   Black-stemmed daisy, *Lagenophora petiolata*, is a spreading plant that forms patches of rosettes of coarsely serrated, orbicular leaves 10–20 mm across, on petioles 2–3 cm long, at the nodes. Flowers, about 8 mm across, arise on long black stalks from November to February. Found throughout New Zealand and the Kermadec Islands in open places in grasslands and shrublands. ASTERACEAE

954   Black-stemmed daisy plant in flower, Ruahine Range (February)

HERBFIELDS AND FELLFIELDS

Herbfields occur in the upper belt of the subalpine region, tending to merge below into the tussocklands and above into the fellfields. The herbfields form extensive areas on the more gentle slopes in the mountains and are characterised by an abundance of tall and medium-sized herbs, sometimes intermingled with tussocks and grasses and an occasional shrub. Fellfields are steeper stony areas with little soil and extend above the herbfields to the permanent snow-line. Fellfield plants are mostly low-growing species, forming more open associations among the rocks and capable of surviving on poor soils in a harsh climate.

955–956 Giant buttercup/Mt Cook lily, *Ranunculus lyallii*, is the largest ranunculus plant in the world (fig. 955), attaining a height of 1–1.5 m with leaves 13–20 cm across. Found in herbfields, especially alongside creeks and streams, from Marlborough to Fiordland, it produces large panicles of white flowers (fig. 956), each 4–5 cm across, from October till January. Although common, it is local in occurrence but makes an unforgettable sight when found clothing an entire slope with these sparkling flowers.

RANUNCULACEAE

955   Giant buttercup in flower, Homer Cirque (December)

956   Close-up of flowers of giant buttercup, Homer Cirque (December)

957   Lobe-leaf buttercup flowers and leaves, Mt Holdsworth (December)

957   Lobe-leaf buttercup, *Ranunculus clivalis*, is a slender, lax but erect buttercup with a few deeply lobed leaves, 5–10 cm across, on slender petioles 7–15 cm long. Flowers, 2–3 cm across, occur sparingly on stalks up to 60 cm high during December and January. Found in wet situations in subalpine grasslands and herbfields and shrubland where the delicate stems are supported by surrounding vegetation, from the Ruahine Range south to the mountains of north-west Nelson.   RANUNCULACEAE

958   Creeping matipo, *Myrsine nummularia*, is a prostrate, rambling and trailing shrub, found among herbfields, fellfields and along scree edges throughout New Zealand. Flowers occur from October till February, and the berries mature during April and May.   MYRSINACEAE

958   Creeping matipo spray with berries, Arthur's Pass (April)

959   Mountain mikimiki, *Cyathodes empetrifolia*, is a prostrate shrub with wiry branches up to 40 cm long and hairy branchlets, to 15 cm high, bearing thick, narrow leaves, 3–5 mm long, with recurved margins. Flowers occur from November to February and the red fruits, 3–5 mm long, from January till April. Found in subalpine herbfields, fellfields, grasslands and boggy places throughout New Zealand.   EPACRIDACEAE

959   Mountain mikimiki plant with fruits, Volcanic Plateau (February)

960   Yellow rock daisy, *Brachyglottis lagopus*, is a small herb clothed with hairs and recognised by its soft reticulated leaves, 3–15 cm long by 3–10 cm wide, on stout hairy petioles, 3–10 cm long. Flowers, 2–4 cm across, on hairy stems up to 35 cm high, occur from January to March, and the plant is found in herbfields, fellfields and grasslands, mostly in the shade of other plants, from the Ruahine Range to Otago.   ASTERACEAE

960   Yellow rock daisy plant in flower, Mt Holdsworth (February)

961-962 Mountain snowberry, *Gaultheria depressa* var. *novae-zelandiae*, forms a prostrate, creeping, rooting shrub with interlacing branches and hairy branchlets bearing thick, leathery, crenulate leaves, 5-10 mm long by 4-6 mm wide, on short petioles (fig. 961). Flowers start in November and continue till February, followed by white or pink berries (fig. 962), 3-4 mm across, from January onwards. Found in subalpine and alpine herbfields and fellfields, grasslands, boggy and rocky places from the Kaimanawa Mountains to Stewart Island.

ERICACEAE

963 Alpine groundsel, *Senecio bellidioides*, is a small plant with rounded, rugose, leathery leaves, up to 5 cm long, on hairy petioles with, usually, hairy margins; the leaves are often appressed to the ground. Flowers, 2-3 cm across, arise on black hairy stalks, up to 30 cm long, from November till January and the plant is found in subalpine herbfields and grasslands throughout the South Island.

ASTERACEAE

963 Alpine groundsel plant in flower, Lewis Pass (December)

961 Mountain snowberry plant in full flower, Homer Cirque (December)

962 Berries of mountain snowberry, Lewis Pass (January)

964 Brockie's bluebell, *Wahlenbergia brockiei*, forms clumps of leafy rosettes crowded together; the leaves are 10-30 mm long by 1-2 mm wide. Flowers, 16-20 mm across, arise on slender stems, to 10 cm high, from each rosette from November to January. Found growing on limestone soils near Castle Hill.

CAMPANULACEAE

964 Brockie's bluebell plant in flower, Otari (January)

965   Alpine avens in flower, Mt
Lucretia (January)

965 Alpine avens, *Geum uniflorum*, is a herbaceous plant, forming large patches in damp areas of subalpine and alpine herbfields, fellfields and open rocky places from the Nelson mountains to Otago. The rounded, hairy-margined leaves arise directly from the root stock, and flowers, to 2.5 cm across, occur during January and February.

ROSACEAE

966   Mountain heath, showing flowers and leaves, Sugarloaf, Cass (November)

966–967 Patotara/mountain heath, *Leucopogon suaveolens*, is a prostrate, branching shrub, forming large patches or dense hummocks, 8–15 cm high and up to 1 m across, in herbfields, fellfields, grasslands and exposed places from the Volcanic Plateau southwards. The leaves, 5–9 mm long, are characterised by five parallel veins on each underside, and the leaf margins are hairy. Flowers, each about 7 mm long, arise as 2–5-flowered terminal racemes from November to February. The fruit, about 3 mm across, may be white, pink or crimson and occurs from January to April.   EPACRIDACEAE

967   Mountain heath, spray showing fruit and leaves, Volcanic Plateau (February)

968 Tall pinatoro, *Pimelea buxifolia*, is a shrub, to 1 m high, found in fellfields and alpine grasslands among the North Island mountains. Leaves, 5–10 mm long by 3–5 mm wide, are keeled, and flowers occur from September to April. A very similar plant, *P. traversii*, is found in fellfields throughout the South Island mountains.

THYMELAEACEAE

968   Spray of tall pinatoro, showing flowers and leaves, Mt Ruapehu (December)

969–970 Mountain pinatoro, *Pimelea oreophila,* is a small shrub, 10–15 cm high, with ascending branches; the older branches are scarred by rings left by falling leaves, the younger branches are hairy. Leaves, 3–6 mm long by 1–3 mm wide, have hairy margins, and the sweet-scented flowers, 6 mm across, which appear from October to March, give rise to white, ovoid berries, 2 mm across, from January on. Found in subalpine herbfields, grasslands and exposed places throughout New Zealand.

THYMELAEACEAE

971–972 Common drapetes, *Drapetes dieffenbachii,* is a sprawling, prostrate plant, forming patches up to 30 cm across in subalpine herbfields, fellfields and grasslands from the Coromandel Ranges south to Stewart Island. Leaves, 2.5–3.5 mm long, are appressed to the branches and taper from base to apex. Clusters of flowers occur terminally on the branches from November to January.

THYMELAEACEAE

971 Common drapetes plant in flower, Mt Holdsworth (December)

969 Mountain pinatoro spray, close up, showing flowers, leaves and branch, Sugarloaf, Cass (November)

970 Mountain pinatoro flowers, Sugarloaf, Cass (November)

972 Flowers of common drapetes, close up, Mt Holdsworth (December)

973-974 Matted ourisia, *Ourisia vulcanica*, is a spreading and rooting herb, forming a mat 10-15 cm across. Its thick, fleshy, hairless leaves, 10-25 mm long by 6-15 mm wide, are appressed to the ground, and flowers, 15-20 mm across, on hairy stalks up to 12 cm high, occur from October to January. Found in dry, sunny places on herbfields, fellfields and exposed places on the Kaimanawa Ranges and the Volcanic Plateau. SCROPHULARIACEAE

973　Matted ourisia flowers and flower-buds, Mt Ruapehu (December)

974　Matted ourisia plant with flowers, Mt Ruapehu (December)

Mountain daisies

Daisies belonging to the genus *Celmisia* are among the most common plants found in the New Zealand mountains. Some have already been met with among the tussocklands but the most species by far are found on the herbfields and fellfields of our mountains. A selection of these is illustrated here.

ASTERACEAE

975-976 Brown mountain daisy, *Celmisia traversii*, is the only *Celmisia* with a rich brown-coloured, velvety tomentum on its leaves and stems (fig. 976). The plant (fig. 975) occurs in herbfields from north-west Nelson south to the Lewis Pass, and flowers, 4-5 cm across, occur during December and January.

975　Brown mountain daisy plant in flower, Mt Lucretia (January)

976　Close-up of leaves of brown mountain daisy, Mt Lucretia (January)

977 Strap-leaved daisy plant in flower, Fog Peak (January)

977 Strap-leaved daisy, *Celmisia angustifolia*, is a small, woody plant with leathery, strap-like leaves, 3–5 cm long, flowering from December to February, and found in subalpine fellfields and herbfields from Arthur's Pass to the Humboldt Mountains.

980–981 Large mountain daisy/tikumu/silvery cotton plant, *Celmisia coriacea*, forms a large, tufted herb, with leaves to 60 cm long, silvery above when young but green when old (fig. 981). The large, symmetrical flower (fig. 980), 5–12 cm across, occurs on fluffy, woolly stems, to 75 cm high, from December to February. Found in alpine herbfields, fellfields and grasslands throughout the South Island.

978–979 White cushion daisy, *Celmisia sessiliflora*, forms a spreading perennial herb with rigid leaves in tight rosettes, which together can form a mat (fig. 978) 1 m across. Sessile flowers, 10–20 mm across (fig. 979), occur during December and January. Found throughout the South Island in subalpine and alpine herbfields and fellfields.

978 White cushion daisy plant in flower, Wapiti Lake, Fiordland (December)

979 Close-up of flower of white cushion daisy, Mt Robert (January)

980 Large mountain daisy, showing the symmetry of the flower and flower-bud, Arthur's Pass (January)

981 Large mountain daisy plants in full flower, Jack's Pass (January)

982   Armstrong's daisy, plants in full flower, Arthur's Pass (January)

982–983   Armstrong's daisy, *Celmisia armstrongi* (fig. 982), and **Lance-leaved daisy,** *Celmisia lanceolata* (fig. 983), are both tufted herbs with rigid, lance-like leaves up to 35 cm long; those of *lanceolata* each have a distinct, stout, yellowish orange-coloured midrib; those of *armstrongi* each have a yellowish orange band down each side of the midrib. Both flower during January and February and are found in herbfields and fellfields among the mountains of the South Island.

984   Sticky-stalked daisy, *Celmisia hieracifolia*, is a grassland to herbfield daisy found from the Ruahine and Tararua Ranges south to the mountains of Nelson. Flowers, 3–4 cm across, occur on sticky, hairy stalks during January and February.

983   Lance-leaved daisy plants with flowers, Stillwater Basin, Fiordland (December)

984   Sticky-stalked daisy in flower, Brown Cow Pass, Boulder Lake (February)

985   Haast's daisy, *Celmisia haastii*, forms patches to 75 cm across; the leaves, 4–7 cm long, are leathery, green above with distinct longitudinal grooves and with satiny hairs below. Flowers, up to 4 cm across, occur during December and January. The whole plant is very sticky and is found throughout South Island mountains in herbfields and fellfields.

985   Haast's daisy plants in flower, Temple Basin (January)

986   Purple-stalked daisy, *Celmisia petiolata*, forms a large, tufted herb with smooth, silky leaves, 7–15 cm long, having purple lateral veins and flat, purple petioles. Flowers, 5–7 cm across, occur on purple, hairy stalks, 15–20 cm high, during December and January. Found in western herbfields and fell-fields from the Spencer Mountains southwards.

986   Purple-stalked daisy in flower, Arthur's Pass (January)

987   Boulder Lake daisy, *Celmisia parva*, is found only near Boulder Lake and on the Paparoa Range in herbfields and grasslands where the climate tends to be wetter. The leaves, 5–15 cm long by 10–15 mm wide, each tapering to its apex, have recurved, faintly toothed margins and dense appressed hairs below. Flowers, 2–3 cm across, occur during January and February.

987   Boulder Lake daisy in flower, Douglas Ridge, Boulder Lake (February)

988   Pigmy daisy, *Celmisia lateralis*, is a sprawling plant with branching stems 30 cm long and thick, dense, overlapping, hairless leaves, 6–8 mm long by 1–2 mm wide. Flowers, 10–20 mm across, occur on 8 cm long, slender stalks during January and February. Found only in herbfields and fellfields of the mountains of north-west Nelson and the Paparoa Range.

988   Pigmy daisy plant with flowers, Lead Hill (February)

989   Dagger-leaf daisy, *Celmisia petriei*, is a tufted daisy with rigid, thick, sharply pointed, dagger-like leaves, 20–30 cm long by 10–20 mm wide, bright green above with two prominent ridges. Flowers, 3–4 cm across, occur on stout, woolly stems, 20–50 cm high, from December to February. Found in herb- and fellfields from the Nelson mountains to Fiordland.

989   Dagger-leaf daisy in flower, Douglas Ridge, Boulder Lake (February)

990   Cotton daisy/cotton plant, tikumu, *Celmisia spectabilis*, forms a stout, rosulate, tufted herb, sometimes forming patches. Leaves, 10–15 cm long by 10–25 mm wide, are very thick and leathery, with the upper surface smooth and shining but the lower surface densely clothed by soft, matted, buff or white hairs. Flowers, 4–5 cm across, on white woolly stalks, 8–25 cm high, occur from December to February. Found in subalpine grasslands, herbfields and fellfields from Mt Hikurangi south to North Otago.

990   Cotton daisy plant in flower, Porter's Pass (January)

991   Allan's daisy, *Celmisia allanii*, forms a loosely branched plant with leaf remains on the branches. The leaves, 3–4 cm long by 10–15 mm wide, are thin, flexible and densely clothed all over with soft white hairs. Flowers, 3–4 cm across, occur on hairy stalks during January and February. Found in herbfields and grasslands from the Nelson mountains to the Lewis Pass.

991   Allan's daisy plant in flower, Mt Lucretia (January)

992   Dainty daisy, *Celmisia gracilenta*, is a slender, tufted daisy with tough, inflexible leaves, 10–15 cm long by 2–4 mm wide, each tapering to a sharp apex and with recurved margins, rolled almost to the midrib, to cover the lower surface of appressed, satiny hairs. Flowers, 10–20 mm across, on slender stalks occur from November to February. Found all over New Zealand in herbfields and subalpine or alpine bogs and grasslands from the Coromandel Peninsula southwards. I have also found it growing splendidly beside the sea at Wharariki Beach at the base of Farewell Spit (fig. 1008).

992   Flowers and leaves of dainty daisy, Mt Tongariro (November)

993   Marlborough daisy, *Celmisia monroi*, grows as a single plant or as a mass of branching stock, with rigid leaves, 7–15 cm long by 5–20 mm wide, the upper surface grooved longitudinally and covered by a thin, silvery skin. Flowers, 4 cm across, occur during December and January. Found in subalpine herbfields and fellfields of Marlborough and at the Woodside Gorge.

994   Larch-leaf daisy, *Celmisia laricifolia*, is a mat-forming plant with branches clothed by leaf remains and bearing tufts of narrow, pointed, silvery green, larch-like leaves, 10–15 mm long by 1–1.5 mm wide, with the margins rolled almost to the midrib. Flowers, 10–20 mm across, occur during January and February. Found in fellfields and exposed places throughout the mountains of the South Island.

993 Marlborough daisy plant in flower, Woodside Gorge (December)

994 Larch-leaf daisy plant in flower, Temple Basin (January)

995 Sticky daisy plant in flower, Mt Torlesse Range (January)

995 Sticky daisy, *Celmisia viscosa*, forms patches up to 1 m across in herb and fellfields and grasslands of the eastern slopes of the South Island mountains. the leaves, 6–8 cm long by 6–9 mm wide, are moderately sticky, and flowers, 2–4 cm across, occur on sticky stalks, 15–30 cm high, from December to February.

996 Mt Taranaki daisy, *Celmisia major* var. *brevis*, is found in herb- and fellfields on Mt Taranaki. The thick, leathery leaves taper to an acute apex with distinct midrib, recurved margins and satiny hairs below. Flowers, 2.5–3 cm across, occur from December to February.

996 Mt Taranaki daisy plants in flower, Mt Taranaki (December)

997 Downy daisy plants in flower, Mt Taranaki (January)

997 Downy daisy, *Celmisia glandulosa*, is a creeping, rooting plant, forming rosettes of leaves 15–25 mm long and 10–15 mm wide, covered by many minute glands and having finely serrate margins. Flowers, 2 cm across, arise on slender stalks, 7–12 cm high, during December and January. Found throughout New Zealand in herbfields, fellfields, grasslands and exposed places.

998 Musk daisies, *Celmisia du-rietzii* and *Celmisia discolor*, are both strongly musk-scented, sprawling or prostrate herbs, often forming huge patches in alpine herbfields, fellfields, grasslands and rocky places throughout the South Island mountains. Older sections of branches are covered with dead leaves, the younger sections with rosulate tufts of sticky leaves, 2–4 cm long by 8–12 mm wide; in *discolor* the upper surface is lightly clothed with white hairs, in *du-rietzii* it is bare. Flowers, 2–3 cm across, on stalks 10–15 cm high, occur during December and January.

998 Musk daisies *C. discolor*, in full flower, Arthur's Pass (January)

New Zealand gentians

Nineteen species of gentians are known from the mountain regions of New Zealand. All flower in the late summer or autumn. Their flowers are mostly white and the species are difficult to separate. A selection is illustrated here.   GENTIANACEAE

999 Ridge-stemmed gentian, *Gentiana tereticaulis*, is an erect herb with ridged stems, 24–45 cm high. Flowers, to 2 cm across, are both terminal and subterminal, with the upper cauline leaves sessile. Found in herbfields and fellfields from the Nelson Mountains to Fiordland.

1000 Alpine gentian, *Gentiana patula*, is a sprawling herb with stems rising at their tips to 20–50 cm to bear cymes of flowers, each 2–2.5 cm across, during January and February. Found in subalpine and alpine herbfields and grasslands from the Tararua Ranges southwards.

1001 Townson's gentian, *Gentiana townsonii*, forms a herb up to 30 cm high, with crowded, fleshy, basal leaves. The plant is found in subalpine herbfields and grasslands from north-west Nelson to Arthur's Pass and flowers during January and February.

999 Ridge-stemmed gentian flowers, Cupola Basin (April)

1000 Alpine gentian flowers, Mt Holdsworth (February)

1001 Townson's gentian flowers, Boulder Lake (January)

1002 North-west Nelson gentian plant in flower, Brown Cow Pass, Boulder Lake (February)

1003 Pink gentian flowers, Upper Cobb Valley (February)

1004 Snow gentian, mauve flowers, McKenzie Pass (March)

1002 North-west Nelson gentian, *Gentiana spenceri*, is a herb about 15 cm high, with flowers, about 2 cm across, occurring from February to April as umbels each surrounded by a whorl of 5–7 spathulate leaves. Found in shaded situations in subalpine herbfields among the mountains of north-west Nelson and Westland.

1003 Pink gentian, *Gentiana tenuifolia*, is a perennial with finely ridged, erect stems, up to 40 cm high, bearing masses of pink flowers, 2.5 cm across, during February and March. Found in subalpine herbfields from north-west Nelson to the Lewis Pass.

1004–1005 Snow gentian, *Gentiana matthewsii*, forms a tall gentian, branching and dividing from the base (fig. 1005), and found in subalpine to alpine herbfields and fellfields from the Humboldt Mountains to Fiordland. Flowers, about 2.5 cm across, occur in profusion from January to March and can be white or pale mauve (fig. 1004).

1006 Common New Zealand gentian, *Gentiana bellidifolia*, is a herb with crowded tufts of thick, fleshy, elliptic basal leaves, 10–15 mm long by 5–7 mm wide. Flowers, each about 2 cm across, occur during March and April as terminal cymes of 2–6 flowers each, on several stems 5–15 cm high. Found from Mt Hikurangi to Fiordland in damp places in herbfields and grasslands.

1005 Large plant of snow gentian with white flowers, Gertrude Cirque, Fiordland (January)

1006 Common New Zealand gentian flowers, Arthur's Pass (April)

PLANT ASSOCIATIONS

Most New Zealand plants will grow singly, as isolated specimens, but in their natural state they normally grow in intimate and often complicated associations and communities. Forests, the most complicated and extensive communities, exert a profound effect upon climate and help to minimise land erosion. Along the New Zealand sea coasts and in the mountains delicate, often fragile plant associations also play essential roles in land stabilisation and climate regulation.

1007 Looking north-east from Mt Lucretia over the Lewis Pass showing the characteristic demarcation line between the beech forest and the subalpine and alpine regions (January)

1009 An alpine meadow on Jack's Pass, North Canterbury (January).

1008 An aberrant association of the subalpine *Celmisia gracilenta* growing in sand at Wharariki Beach, Farewell Spit. Other New Zealand *Celmisia* also occur in this aberrant fashion around the Fiordland Sounds.

1010 A rain forest interior in a Westland podocarp forest shows the nature of a primeval forest community in New Zealand, Fox Glacier (January).

GLOSSARY

acuminate: tapering to a fine point

adpressed: pressed closely together but not joined

alveolate: deeply and closely pitted

apetalous: not having any petals

appressed: closely applied to

aril: an outgrowth or appendage to a seed

attenuated: gradually tapering to a point

axillary: the upper angle between two structures

calyx: the outer whorl of the parts of a flower, usually green, and made up of the separate or fused sepals

capsule: a dry fruit that splits at maturity to release its seeds

carpidium: the scale at the base of the cone in Gymnosperms

ciliated: fringed with hairs along the margin

cladodes: flattened stems having the function of leaves

corymb: a flat topped cluster of flowers in which the longer stalked outer flowers open first

costa: a rib, especially the midvein of a leaf

crenate: having rounded shallow teeth

cupule: a cup-like structure occurring at the base of some flowers

cyme: an inflorescence in which the principal stalk ends with a flower and in which other flowers are produced at the ends of lateral stalks

decumbent: lying along the ground but with the tip ascending

dentate: having sharp teeth at right angles to the margin

dioecious: having male and female flowers on different plants

divaricating: spreading at very wide angles

drupe: a fruit with a seed enclosed in a bony cover surrounded by a fleshy layer, a "stone fruit"

emarginate: having a narrow notch at the apex

fascicles: a closely bunched cluster

glabrous: without any hairs

globose, globate: a round-shaped structure

hybrid: an organism produced from interbreeding of two different species

imbricate: overlapping, like the tiles on a roof

inflorescence: an arrangement of flowers or a collection of flowering parts

lanceolate: lance-shaped, tapering from about one third from base towards the apex

leaflet: a single element of a compound leaf

monoecious: having male and female flowers on the same plant

montane: a zone between the lowland and subalpine regions

obovate: egg-shaped

ochreous: dull yellow coloured with often a tinge of red

panicle: a branching inflorescence of flowers with each flower on a stalk

pedicel: the stalk that holds a single flower

peduncle: the stalk of a solitary flower or the stalk of a compound flower head *perianth:* the outer whorl of a flower when it is not distinctly formed from sepals and petals; a general term for the calyx and the corolla together

petiole: the stalk of a leaf

pinna: a primary division of a divided leaf

pinnate: with the parts divided and arranged on either side of the axis: compound

pinnatifid, pinnatisect: compound

raceme: an unbranched elongate inflorescence with flowers having stalks; the flowers at the base being the oldest

rachis: the main axis of an inflorescence; plural is rachides

reticulated: forming a network

revolute: rolled outwards or rolled to the lower side

rhizome: an underground, spreading stem

rosulate: forming a small rosette

rugose: wrinkled

serrated: sharply toothed with teeth pointing forwards or outwards

sessile: without a stalk

spathulate: spoon-shaped

strobilus: a cone-shaped structure as in pines and lycopods; plural is strobili

terminal: borne at the apex of a stem

tomentum: a more or less dense covering of matted or appressed hairs, described as tomentose

trifoliate: having three leaves

trifoliolate: having three leaflets

umbel: a usually flat-topped inflorescence having the pedicels arising from a common centre point

unifoliate: having a single leaf

REFERENCE WORKS

Adams, Nancy M., *Mountain Flowers of New Zealand*. A.H. & A.W. Reed, Wellington, 1965.

Allan, H.H., *Flora of New Zealand: Volume 1*. Government Printer, Wellington, 1961.

Brooker, S.G.; Cambie, R.C.; and Cooper, R.C., *New Zealand Medicinal Plants*. Revised edition, Heinemann, Auckland, 1987.

Cartman, Joe, *Growing New Zealand Alpine Plants*. Reed Methuen, Auckland, 1985.

Cheeseman, T.F., *Manual of the New Zealand Flora*. Government Printer, Wellington, 1925.

Cockayne, L., *New Zealand Plants and Their Story*. Government Printer, Wellington, 1910.

—— *The Vegetation of New Zealand*. Third Edition, Engelmann, Leipzig, 1958.

Cockayne, L. and Turner, E.P., *The Trees of New Zealand*. Second edition, Government Printer, Wellington, 1958.

Connor, H.E., *The Poisonous Plants in New Zealand*. New Zealand DSIR Bulletin 99, Second edition, 1977.

Crowe, Andrew, *A Field Guide to the Native Edible Plants of New Zealand*. Collins, Auckland, 1981.

Eagle, A.S., *Eagle's Trees and Shrubs of New Zealand in Colour*. Collins, Auckland, 1975.

—— *Eagle's Trees and Shrubs of New Zealand in Colour*. Second series, Collins, Auckland, 1982.

—— *100 Shrubs and Climbers of New Zealand*. Collins, Auckland, 1978.

Featon, E.H., *The Art Album of New Zealand Flora*. Bock & Cousins, London, 1889.

Fisher, Muriel E.; Satchell, E.; and Watkins, Janet M., *Gardening with New Zealand Plants, Shrubs and Trees*. Collins, Auckland, 1970.

Given, David R., *Rare and Endangered Plants of New Zealand*. A.H. & A.W. Reed, Wellington, 1981.

Healy, A.J. and Edgar, E., *Flora of New Zealand: Volume 3*. Government Printer, Wellington, 1980.

Johnson, Marguerite, *New Zealand Flowering Plants*. Caxton Press, 1968.

Johnson, Peter and Brooke, Pat, *Wetland Plants in New Zealand*. DSIR Publishing, Wellington, 1989.

Laing, R.M. and Blackwell, E.W., *Plants of New Zealand*, Fourth edition, Whitcombe and Tombs Ltd., Wellington, 1940.

Malcolm, Bill and Nancy, *New Zealand's Alpine Plants Inside and Out*. Craig Potton, Nelson, 1988.

Mark, A.F. and Adams, N.M., *New Zealand Alpine Plants*. A.H. & A.W. Reed, Wellington, 1973.

Metcalf, L.J. *The Cultivation of New Zealand Trees and Shrubs*. A.H. & A.W. Reed, Wellington, 1972.

Moore, L.B., and Adams, Nancy M., *Plants of the New Zealand Coast*. Pauls Book Arcade, Auckland, 1963.

Moore, L.B. and Edgar, E., *Flora of New Zealand: Volume 2*. Government Printer, Wellington, 1970.

Moore, L.B. and Irwin, J.B., *The Oxford Book of New Zealand Plants*. Oxford University Press, Wellington, 1978.

Philipson, W.R. and Hearn, D., *Rock Garden Plants of the Southern Alps*. Caxton Press, Christchurch, 1962.

Poole, A.L. and Adams, N.M., *Trees and Shrubs of New Zealand*. Revised edition, Government Printer, Wellington, 1979.

Richards, E.C., *Our New Zealand Trees and Flowers*. Third edition, Simpson and Williams Ltd, Christchurch, 1956.

Salmon, J.T., *The Native Trees of New Zealand*. A.H. & A.W. Reed, Wellington, 1980.

—— *Collins Guide to the Alpine Plants of New Zealand*. Collins, Auckland, 1985.

—— *A Field Guide to the Native Trees of New Zealand*. Reed Methuen, Auckland, 1986.

Sampson, F. Bruce, *Early New Zealand Botanical Art*. Reed Methuen, Auckland, 1985.

Webb, Colin; Johnson, Peter; and Sykes, Bill, *Flowering Plants of New Zealand*. DSIR, Christchurch, 1990.

Wilson, Catherine M. and Given, David R., *Threatened Plants of New Zealand*. DSIR Publishing, Wellington, 1989.

Wilson, H.D., *Field Guide, Wild Plants of Mount Cook National Park*. Field Guide Publications, Christchurch, 1978.

—— *Field Guide, Stewart Island Plants*. Field Guide Publications, Christchurch, 1982.

INDEX